Statistical Graphics in SAS®

An Introduction to the Graph Template Language *and the* Statistical Graphics Procedures

Warren F. Kuhfeld

The correct bibliographic citation for this manual is as follows: Kuhfeld, Warren F. 2010. *Statistical Graphics in SAS®: An Introduction to the Graph Template Language and the Statistical Graphics Procedures.* Cary, NC: SAS Institute Inc.

Statistical Graphics in SAS®: An Introduction to the Graph Template Language and the Statistical Graphics Procedures

Contents

Preface

Effective graphics are indispensable for modern statistical analysis. They reveal patterns, differences, and uncertainty that are not readily apparent in tabular output. Graphics provoke questions that stimulate deeper investigation, and they add visual clarity and rich content to reports and presentations.

There are three major ways in which you can use statistical graphics:

- **Exploring your data.** Many statistical analyses are begun by producing simple graphs of data such as scatter plots, histograms, and box plots.

- **Analyzing your data.** Graphical results are an integral part of the results of modern statistical analyses.

- **Presenting your data.** The final results of statistical analyses are often presented with graphs. Presentation graphics might be the same as the graphical results of analyses or they might be customized in various ways.

The SAS System provides many tools for creating statistical graphics. In many cases, multiple tools can be used to make a graph.

- **Automatically created graphs.** With SAS 9.2, statistical graphs are created with an extension of the Output Delivery System (ODS). ODS manages procedure output and lets you display it in a variety of destinations, such as HTML and RTF. With ODS Statistical Graphics (or ODS Graphics for short), statistical procedures produce graphs as automatically as they produce tables, and graphs are integrated with tables in the ODS output. You only need to enable ODS Graphics with the `ods graphics on` statement to get default graphical output. Some graphs require you to additionally specify one or more simple options.

- **SG Procedures.** The SG (Statistical Graphics) procedures SGPLOT, SGSCATTER, and SGPANEL provide a simple and convenient syntax for producing many types of statistical graphs. They are particularly convenient for exploring and presenting data.

- **The GTL.** The GTL (Graph Template Language) and PROC SGRENDER provide a powerful syntax for creating custom graphs. You can also modify the templates that SAS provides for use with SAS/STAT procedures to create customized results.

Table 1 provides a summary of ODS Graphics functionality.

Table 1 Summary of ODS Graphics Functionality

	Graphical Task	Audience	What do you use?	What should you learn?	
✔	Create graphs in the context of statistical analyses	Statistical users	Statistical procedures in SAS/STAT, SAS/ETS, SAS/QC, and Base SAS that support ODS Graphics	Specify ODS GRAPHICS ON; graphs are then created by default or with procedure options which are documented in the procedure chapters	Minimal graph syntax
	Enhance specific graphs for a paper or presentation	Statistical and general SAS users	ODS Graphics Editor	How to request editable graphs, invoke the Editor, use point-and-click features; see the *SAS/GRAPH: ODS Graphics Editor User's Guide*	
✔	Create stand alone graphs for data exploration or for customized displays	Statistical and general SAS users	SGPLOT, SGPANEL, SGSCATTER procedures in SAS/GRAPH	SG procedure syntax; see the *SAS/GRAPH: Statistical Graphics Procedures Guide*	Some syntax
✔	Change the overall consistent appearance of graphs and tables	Statistical and general SAS users	ODS styles	STYLE= option in ODS destination statement	
✔	Save and manage graphs for papers and presentations	Statistical and general SAS users	ODS GRAPHICS options, ODS destination options	How to specify size and resolution; how to name and access image files	
✔	Make persistent changes in graphs produced by statistical procedures (apply whenever you run your program)	Advanced SAS programmers	User-modifications of graph templates that SAS provides	Basic features of the Graph Template Language and PROC TEMPLATE; see the *SAS/GRAPH Template Language Reference*	Graphics programming
	Create a highly customized stand alone graph	Advanced SAS programmers	ODS Graphics Designer	GUI for creating graph templates	
✔	Create a highly customized stand alone graph	Advanced SAS programmers	User-written graph templates	Graph Template Language, PROC TEMPLATE, and PROC SGRENDER; see the *SAS/GRAPH Template Language Reference* and the *SAS/GRAPH: Graph Template Language User's Guide*	

✔ discussed in this book

Figure 1 Fit Plot Created by PROC GLM

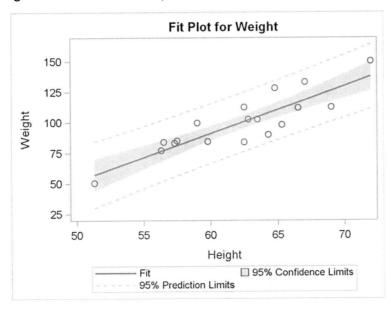

Figure 2 ANOVA and Fit Statistics

```
                        The GLM Procedure

Dependent Variable: Weight

                                 Sum of
        Source            DF     Squares     Mean Square   F Value   Pr > F

        Model              1   7193.249119   7193.249119    57.08    <.0001

        Error             17   2142.487723    126.028690

        Corrected Total   18   9335.736842


              R-Square    Coeff Var     Root MSE     Weight Mean

              0.770507    11.22330      11.22625      100.0263
```

The graphs that are automatically produced by SAS/STAT procedures are suitable for use in reports. However, you might want to customize them first (for example by using the point-and-click ODS Graphics Editor). (See *SAS/GRAPH: ODS Graphics Editor User's Guide*.) This book focuses primarily on the GTL and its use in creating customized graphs. It also presents the SG procedures and shows how to make most graphs in multiple ways. The SG procedures are easier to use than the GTL, but they are less powerful and less flexible. Both simple and complex examples are provided to help you get started creating modern statistical graphics with the GTL and SG procedures.[1]

---

[1] Emphasis is placed on the code and the graphs, not on the data or the results.

Figure 3 Fit Plot Created by PROC SGPLOT

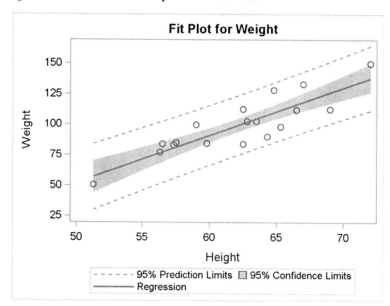

ODS Graphics is available in procedures in SAS/STAT, Base SAS, SAS/ETS, SAS/QC, SAS/GRAPH, and other products. SAS/GRAPH software is required for ODS Graphics functionality. The following step creates the regression fit plot displayed in Figure 1 along with the ANOVA table and fit statistics in Figure 2:

```
ods graphics on;

proc glm data=sashelp.class;
   model weight = height;
run;
```

Basic ODS Graphics functionality is discussed in many places including in "A Primer on ODS Statistical Graphics" (*SAS/STAT User's Guide*). The documentation chapter for each SAS/STAT, SAS/QC, Base SAS, and SAS/ETS procedure that uses ODS Graphics provides examples, information about which graphs are produced by each procedure, and information about what syntax (if any) is required to make each plot.

ODS Graphics provides more than just a way to automatically produce graphs from analytical procedures. It also provides two powerful ways to produce custom graphs. First, there are the SG procedures. These procedures are PROC SGPLOT, PROC SGSCATTER, and PROC SGPANEL.[2] They provide a high level syntax for producing scatter plots, histograms, bar charts, box plots, scatter plot matrices, and many other types of statistical graphs. You can use these procedures instead of PROC GPLOT, PROC GCHART, and other legacy SAS/GRAPH procedures. The SG procedures are designed specifically for statistical work, and have statistical computation facilities directly built in to them. For example, the following step produces a scatter plot with a linear regression line, confidence limits, and prediction limits that is very similar to the plot that PROC GLM produces:

---

[2]ODS Graphics does not need to be enabled with the ODS GRAPHICS statement to use the SG procedures or PROC SGRENDER.

```
proc sgplot data=sashelp.class;
   title 'Fit Plot for Weight';
   reg y=weight x=height / cli clm;
run;
```

The results are displayed in Figure 3. The plots in Figure 2 and Figure 4 differ in the degree of transparency of the band that displays the confidence limits and in the order and labels of the entries in the legend.

In addition to the SG procedures, ODS Graphics provides you with the Graph Template Language, PROC TEMPLATE, and PROC SGRENDER.[3]

The GTL is a powerful language for defining the layout and composition of a graph or a panel of graphs. You can use it to make very simple and very complex graphs. It is the same language that SAS procedures use to define their ODS Graphics. SAS procedures that provide ODS Graphics have a template, written in the GTL, for each graph. You, too, can use the GTL and PROC TEMPLATE to create and compile a graph template. Then you can use PROC SGRENDER to produce graphs from a SAS data set using the instructions provided in the graph template. For example, the following step produces a scatter plot with a linear regression line, confidence limits, and prediction limits that is identical to the plot that PROC GLM produces:

```
proc template;
   define statgraph FitPlot;
      begingraph;
         entrytitle 'Fit Plot for Weight';
         layout overlay;
            modelband "cliband" / display=(outline)
                       outlineattrs=GraphPredictionLimits
                       name='cli' legendlabel='95% Prediction Limits'
                       datatransparency=0.5;
            modelband "clmband" /
                       fillattrs=GraphConfidence
                       name='clm' legendlabel='95% Confidence Limits'
                       datatransparency=0.5;
            scatterplot y=weight x=height;
            regressionplot y=weight x=height / name='reg'
                          clm="clmband" cli="cliband" legendlabel='Fit';
            discretelegend 'reg' 'clm' 'cli';
         endlayout;
      endgraph;
   end;
run;


proc sgrender data=sashelp.class template=FitPlot;
run;
```

---

[3]The procedure name "SGRENDER" begins with "SG"; however, the phrase "SG procedures" usually refers only to the SGPLOT, SGSCATTER, and SGPANEL procedures.

Figure 4 Fit Plot Created with the GTL

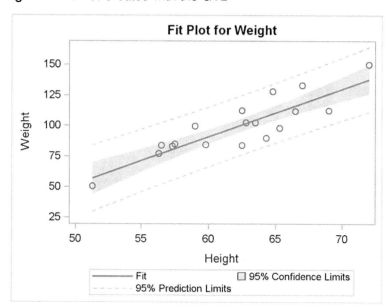

The results are displayed in Figure 4. The graph template statements and options used in this example are explained in detail throughout this book.

These three examples illustrate an important point. In most cases, you can simply enable ODS Graphics and use an analytical procedure to produce your graphs. When you need additional graphs or customized graphs, you can often use an SG procedure to get what you need. However, you can always use the GTL to specify precisely what you need. Using the GTL will almost always be more work, but it gives you the most power. The SG procedures and ODS Graphics are documented in the *SAS/GRAPH: Graph Template Language User's Guide*, the *SAS/GRAPH: Statistical Graphics Procedures Guide*, and the *SAS/GRAPH Template Language Reference*.

This book is organized by type of graph. Alternative ways of creating the same graph are presented together and in parallel. This book provides a gentle, parallel, and example-driven introduction to the GTL and to the SG procedures. Most graphs are produced in at least two ways. One graph is created with the GTL. Whenever possible, the other graph is created more directly with one of the SG procedures. Some graphs are produced by SAS/STAT procedures as well or with additional options. Each example provides prototype code for getting started with the GTL and with the SG procedures. Examples of all of the basic graphs that can be produced by the SGPLOT, SGSCATTER, and SGPANEL procedures are provided. While you do not need to write a template to make many useful graphs, understanding the GTL enables you to create custom graphs that cannot be produced by the SG procedures. It also helps you modify the sometimes complex templates that SAS provides.

Examples of all of the major statements in the GTL are provided. These statements can be classified as follows:

- Plot statements specify a number of commonly used displays, including scatter plots, histograms, contour plots, surface plots, and box plots.
- Layout statements specify the arrangement of the components of the graph.
- Text statements specify the descriptions that accompany graph elements.
- Control statements specify the conditional or iterative flow of control.

This book is about graphics, not statistics. Statistical graphics are created and displayed, but the book does not talk about interpretation or the complexities of data analysis. It uses simple data sets, often the Sashelp.Class data set that is familiar to most SAS users. Other data sets are used as needed. The goal is to show you how to make graphs, not explain why you would do it or what you would do afterward.

The Role of the GTL in SAS

It is important to understand that the GTL is a language developed for internal use at SAS. SAS procedure writers use the GTL to make the graphs that automatically come out of analytical procedures. You can use the same GTL to make your graphs. The GTL was developed to be a comprehensive language for capturing the definition of potentially very complex graphs. While SAS makes this language available to you, it is not designed to be an obvious step away from syntax that is familiar to long-time SAS users. It is a new and different language. That said, you will find as you become experienced with the GTL that it is not hard to use. You will simply find that it differs from what you might expect if you are an experienced SAS user. This is discussed in more detail in Appendix B, "Some Tips and Techniques for Understanding the GTL." If you find that your templates are not working the way you think they should, you might find the answer in the appendix.

Navigating This Book

Much of this book assumes that you are already familiar with using ODS Graphics with SAS analytical procedures but are not familiar with the SG procedures or the GTL. If you are not familiar with the ODS Graphics statements, styles, destinations, and so on, see "Introduction to ODS Graphics" on page 187 for an introduction. You can find an even more detailed introduction in "Statistical Graphics Using ODS" (*SAS/STAT User's Guide*).

Appendix B, "Some Tips and Techniques for Understanding the GTL" provides you with some background information on the GTL. It is important that you understand these concepts when you start writing your own templates. Subsequent sections illustrate ways to make different types of graphs and illustrate one or more statements in the GTL and in the SG procedures. The common statements (DEFINE, BEGINGRAPH, LAYOUT OVERLAY, and so on) are explained in detail only in the section "Scatter Plot" on page 2. Subsequent examples assume that you have read this first example. You should also read the section "Regression Fit Plot" on page 6 before proceeding to other examples. Other than that, you do not need to read each section in order. However, if you come across statements that are unfamiliar, you can use the index to find other and earlier examples of them.[4]

---

[4]Exceptions include the DYNAMIC, MVAR, and NMVAR statements, which are discussed in the sections "Dynamic Variables and Graph Template Modification" on page 153, "Histogram" on page 32, "Bubble Plot" on page 56 and other sections.

The following list can help you find the section that illustrates each type of plot:

After you are familiar with several of the basic plot types, you can read more about other topics in the following sections:

Many sections of this book address topics other than the specific type of graph mentioned. The following table lists some additional topics and the section of the book that addresses them:

Table 2 Cross-Reference to Topics Other Than Graphs

Topic	Section
axis labels and short labels, dynamic variable existence evaluation, data objects, title and footnote modification, adding a project or date to a graph, macro variables, using system titles and footnotes	"Changing Titles and Axis Labels Set by Dynamic Variables" on page 162
band plots, series plots, legends	"Regression Fit Plot with Confidence and Prediction Limits" on page 15
cells and cell headers, side bars, marker characters, LAYOUT LATTICE statement properties, common axes and axis labels	"Residual Panel" on page 95
color ramps, style modification	"Contour Plot" on page 70
complex templates	"Conditional Template Logic" on page 175
computed variables, data object	"Legends" on page 170
control white space near the axis, specify ticks, control data region viewed, set axis labels, change marker type and size	"Scatter Plot" on page 2
controlling group level order, shared axes	"Data Panel" on page 106
create an all-color style	"Grouped Regression Fit Plot" on page 13
design height and width, LAYOUT LATTICE statement properties, gutters (space between graphs), using the macro language when writing a template	"Scatter Plot Matrix" on page 86
EVAL function, Boolean expressions, observation selection	"Drop Lines" on page 40
EVAL function, parameterized statement, transparency	"Bar Chart" on page 27
histograms, kernel density estimation	"Density Plot" on page 36
macro variables, tick increment calculations, specifying tick values, control data region viewed, parameterized statement, hidden scatter plot, aspect ratio, MVAR and NMVAR statements	"Bubble Plot" on page 56

Table 2 *continued*

Topic	Section
modifying a template created with a LINK statement	"BY Groups" on page 183
ODS trace output, source for templates that SAS provides, Unicode, Greek letters	"Changing Titles and Axis Labels" on page 159
parameterized lines (diagonal reference line), square plot, axis labels, using a SAS supplied template with PROC SGRENDER	"Dynamic Variables and Graph Template Modification" on page 153
parameterized statement	"Box Plot" on page 44
parameterized statement, macro variables	"Histogram" on page 32
reference lines, axis suppression	"Modifying Colors, Lines, Markers, Axes, and Reference Lines" on page 167
reference lines, line attributes, equated axes	"Vector Plot" on page 68
statement order, primary statement, control data region viewed, change marker type and size	"Contour Plot and Scatter Plot Overlaid" on page 74
Unicode, subscripts, superscripts, Greek letters, $\hat{\mu}$, macro variables, gridded layout	"Text Insets and Special Characters" on page 23
use an SG procedure to write a template, suppress automatic legends	"Regression Fit Plot" on page 6

Acknowledgments

I would like to thank Robert Rodriguez and Robert Derr for their helpful comments on an earlier draft of this book. I would also like to thank my editors George McDaniel and Caroline Brickley. Finally, I would like to acknowledge my colleague Jeff Cartier who sadly passed away shortly after the first draft of this book was written. Jeff wrote much of the documentation for ODS Graphics that SAS/STAT developers such as myself used to learn the GTL.

About This Book

Purpose

This book introduces the Graph Template Language (GTL) and the Statistical Graphics (SG) procedures SGPLOT, SGSCATTER, and SGPANEL. This GTL enables you to make a wide variety of modern statistical graphs using a powerful language. The SG procedures enable you to make a more limited set of graphs using a simpler but less powerful syntax.

Is This Book for You?

If you are a SAS/GRAPH, SAS/STAT, or SAS/ETS user and you need to create graphs, then this book is for you.

Scope of This Book

This book illustrates all of the major statements in the GTL and the SG procedures. It shows how to make a wide variety of statistical graphics. However, it does not discuss the underlying statistical methods, data analysis, or interpretation of results.

Typographical Conventions Used in This Book

This book uses several type styles for presenting information. The following list explains the meaning of the typographical conventions used in this book:

roman	is the standard type style used for most text.
UPPERCASE ROMAN	is used for SAS options, and other SAS language elements when they appear in the text. However, you can enter these elements in your own SAS programs in lowercase, uppercase, or a mixture of the two.
MonoSpace	is used for longer SAS options, statements, and statement fragments when they appear in the text. Text is typically lowercase or mixed case. However, you can enter these elements in your own SAS programs in lowercase, uppercase, or a mixture of the two.
oblique	is used for user-supplied values for options in the syntax definitions. In the text, these values are written in *italic*.

helvetica	is used for the names of variables and data sets when they appear in the text.
monospace	is used for example code. This book typically uses lowercase or mixed case type for SAS code.

Software Used to Develop This Book

- The GTL and SG procedures are part of SAS/GRAPH software.

- Many of the examples use SAS/STAT software to prepare data sets for display.

Data and Programs Used in This Book

The data and programs used in this book are available from this book's companion Web site: http://support.sas.com/publishing/authors/kuhfeld.html.

Author Pages

Each SAS Press author has an author page, which includes several features that relate to the author, including a biography and book descriptions for coming soon titles and other titles by the author. Other features include contact information, links to sample chapters and example code and data, events, and occasional extras.

You can access the author pages from http://support.sas.com/publishing/authors.

Example Code—Examples from This Book at Your Fingertips

You can access the example programs for this book by accessing the author pages at http://support.sas.com/authors. Select the author to display the appropriate author page, select the appropriate book, and then click Example Code and Data. For an alphabetical listing of all books for which example code is available, see http://support.sas.com/bookcode. Select a title to display the book's example code. If you are unable to access the code through the Web site, send e-mail to **saspress@sas.com**.

Additional Resources

SAS offers you a rich variety of resources to help build your SAS skills and explore and apply the full power of SAS software. Whether you are in a professional or academic setting, we have learning products that can help you maximize your investment in SAS.

Bookstore	`http://support.sas.com/publishing/`
Training	`http://support.sas.com/training/`
Certification	`http://support.sas.com/certify/`
Knowledge Base	`http://support.sas.com/resources/`
Support	`http://support.sas.com/techsup/`
Learning Center	`http://support.sas.com/learn/`
Community	`http://support.sas.com/community/`

Comments or Questions?

If you have comments or questions about this book, you may contact the author through SAS as follows:

Mail:

SAS Institute Inc.
SAS Press
Attn: Warren F. Kuhfeld
SAS Campus Drive
Cary, NC 27513

E-mail: `saspress@sas.com`
Fax: (919) 677-4444

Please include the title of the book in your correspondence.

For a complete list of books available through SAS Press, visit `http://support.sas.com/publishing`.

SAS Publishing News:

Receive up-to-date information about all new SAS publications via e-mail by subscribing to the SAS Publishing News monthly eNewsletter. Visit `http://support.sas.com/subscribe`.

Chapter 1

GTL and the SG Procedures

Contents

Figure 1.1   Scatter Plot with Custom Template

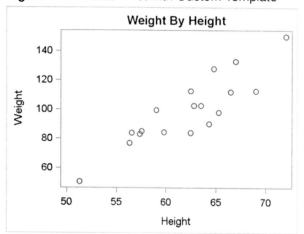

Figure 1.2   Scatter Plot with PROC SGPLOT

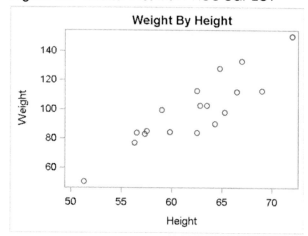

This chapter provides an introduction to the GTL (Graph Template Language) and the SG (Statistical Graphics) procedures. Sections are devoted to different types of plots beginning with sections on the scatter plot and fit plots. The first few examples provide much greater detail than the later examples, so you should read the first few examples before moving on to subsequent examples. Specifically, the common statements (DEFINE, BEGINGRAPH, LAYOUT OVERLAY, and so on) are explained in detail only in the section "Scatter Plot" on page 2. Subsequent examples assume that you have read this first example. You should also read the section "Regression Fit Plot" on page 6 before proceeding to other examples. Other than that, you do not need to read each section in order. However, if you come across statements that are unfamiliar, you can use the index to find other and earlier examples of them. The earlier examples go into more detail about the basics, and the later examples reveal more of the nuances and power of the GTL and the SG procedures.

1.1   Scatter Plot

A scatter plot is a graphical display of two quantitative variables using Cartesian coordinates. The data are displayed as a collection of points, each having the value of one variable on the horizontal axis and the value of the other variable on the vertical axis. The following step creates a graph template that can be used to create a scatter plot of the Sashelp.Class data set:

```
proc template;
   define statgraph classscatter;
      begingraph;
         entrytitle 'Weight By Height';
         layout overlay;
            scatterplot y=weight x=height;
         endlayout;
      endgraph;
   end;
run;
```

A PROC TEMPLATE step for defining a new template begins with a PROC TEMPLATE statement followed by one or more blocks of statements beginning with a DEFINE statement and ending with an END statement. The final RUN statement is not required. Each template is compiled when SAS encounters the END statement that matches the DEFINE statement.

This template, like all graph templates, begins with a DEFINE STATGRAPH statement followed by a template name. In this case, a graph template called `ClassScatter` is created. The next statement in this example is a BEGINGRAPH statement. It provides a place for you to specify options that affect the graph size, the graph border, and the graph background color. Most statements that construct the graph are specified inside the BEGINGRAPH/ENDGRAPH block.[1] Graph titles are specified in the GTL on an ENTRYTITLE statement, which follows the BEGINGRAPH statement. System titles (those specified on TITLE and TITLE*n* statements) and system footnotes do not appear as part of the graph.

For single graphs (as opposed to panels with two or more graphs), the next statement is typically a LAYOUT OVERLAY statement. It is accompanied by an ENDLAYOUT statement, and the statements that create the graph are in between. The LAYOUT OVERLAY statement provides a place for you to specify options that control the ticks, tick labels, axes, axis type, axis labels, grids, and so on. In this case, no options are specified.

The only graph statement is a SCATTERPLOT statement that plots the variable Weight on the Y axis and the variable Height on the X axis. You can use this template to create the scatter plot with the following step:

```
proc sgrender data=sashelp.class template=classscatter;
run;
```

The PROC SGRENDER step consists of simply a data set specification and a template specification. All of the rest of the instructions are in the template. The results are displayed in Figure 1.1.

Alternatively, you can make this plot with PROC SGPLOT and without creating a template as follows:

```
proc sgplot data=sashelp.class;
   title 'Weight By Height';
   scatter y=weight x=height;
run;
```

The input SAS data set is specified in the DATA= option, and the title is specified in the TITLE statement. The SCATTER statement is used to construct the plot with the variable Weight on the Y axis and the variable Height on the X axis. The results are displayed in Figure 1.2. The two plots are identical. When you can use a simple SG procedure step to make the graph you want, you should do so. The GTL is more involved for simpler graphs, but it enables you do make more complex graphs that you cannot make with the SG procedures. Simple GTL examples are provided here to help you create more complex graphs.

---

[1]Exceptions include the DYNAMIC, MVAR, and NMVAR statements, which are discussed in the sections "Dynamic Variables and Graph Template Modification" on page 153, "Histogram" on page 32, "Bubble Plot" on page 56, and other sections.

Figure 1.3 Scatter Plot with Custom Template

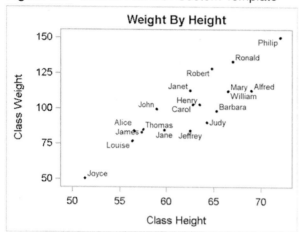

Figure 1.4 Scatter Plot with PROC SGPLOT

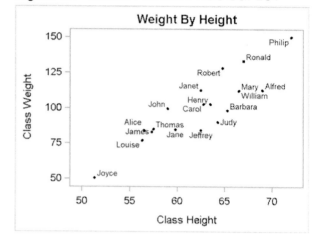

The following steps create a scatter plot while changing the default axes and other options:

```
proc template;
   define statgraph classscatter;
      begingraph;
         entrytitle 'Weight By Height';
         layout overlay /
            xaxisopts=(offsetmin=0.05 offsetmax=0.05 label='Class Height')
            yaxisopts=(offsetmin=0.05 offsetmax=0.05 label='Class Weight'
                        linearopts=(tickvaluesequence=(start=50
                        end=150 increment=25) viewmin=50));
            scatterplot y=weight x=height / datalabel=name
                        markerattrs=(symbol=circlefilled
                                    color=black size=3px);
         endlayout;
      endgraph;
   end;
run;

proc sgrender data=sashelp.class template=classscatter;
run;

proc sgplot data=sashelp.class;
   title 'Weight By Height';
   scatter y=weight x=height / datalabel=name
                              markerattrs=(symbol=circlefilled
                                          color=black size=3px);
   xaxis offsetmin=0.05 offsetmax=0.05 label='Class Height';
   yaxis offsetmin=0.05 offsetmax=0.05 label='Class Weight'
         values=(50 to 150 by 25);
run;
```

Axis options are specified in the LAYOUT OVERLAY statement in the GTL and in the XAXIS and YAXIS statement in PROC SGPLOT. In both plots and on both axes, offsets are specified so that the first 5% (OFFSETMIN=0.05) and the last 5% (OFFSETMAX=0.05) of each axis is left blank. Aesthetically, it is often nice to have a small amount of separation between the axes and the points in the plot. In both plots, an axis label is specified for the X and Y axes by using the LABEL= option. In the GTL, the tick values are specified using these options: `linearopts=(tickvaluesequence=(start=50 end=150 increment=25) viewmin=50))`. The option name LINEAROPTS= is due to the fact that options for log axes are specified differently from options for ordinary linear axes. This set of options produces ticks from 50 to 150 by 25. However, ODS Graphics does not automatically use all of the ticks if they are too far outside the range of the data. The option VIEWMIN=50 ensures that the smallest tick is displayed. Without this option, the first tick is not displayed for these data. There is also a VIEWMAX= option that is not used in this example. In the SCATTER statement in PROC SGPLOT, the same ticks are requested with the option VALUES=(50 TO 150 BY 25). The option DATALABEL=Name uses the Name variable to label the points. The `markerattrs=(symbol=circlefilled color=black size=3px)` option specifies black filled circle markers, three pixels big. The results are displayed in Figure 1.3 and Figure 1.4. The two plots are identical.

The next section starts with a scatter plot and adds to it regression fit functions.

Figure 1.5 Regression with Custom Template

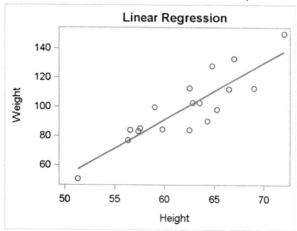

Figure 1.6 Regression with PROC SGPLOT

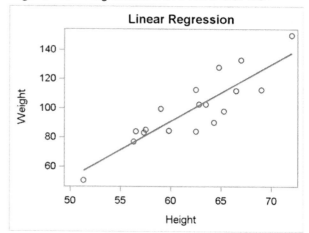

Figure 1.7 Cubic Fit with Custom Template

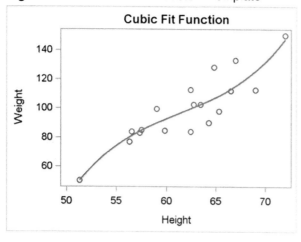

Figure 1.8 Cubic Fit with PROC SGPLOT

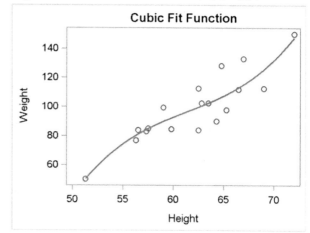

1.2 Regression Fit Plot

A regression fit plot consists of a scatter plot of two quantitative variables along with an overlaid linear or nonlinear fit function. The values on the Y axis are typically computed from the X and Y variables using the method of least squares, although other methods are illustrated in later examples. The following step shows a template that can be used to create a scatter plot of the Sashelp.Class data set along with a linear regression fit line:

```
proc template;
   define statgraph classreg;
      begingraph;
         entrytitle 'Linear Regression';
         layout overlay;
            scatterplot y=weight x=height;
            regressionplot y=weight x=height;
         endlayout;
      endgraph;
   end;
run;
```

The SCATTERPLOT statement plots the variable Weight on the Y axis and the variable Height on the X axis. The REGRESSIONPLOT statement by default fits a line through the points defined by the variable Weight on the Y axis and the variable Height on the X axis. You can use this template and create the plot with the following step:

```
proc sgrender data=sashelp.class template=classreg;
run;
```

The results are displayed in Figure 1.5.

Alternatively, you can make this plot with PROC SGPLOT as follows:

```
proc sgplot data=sashelp.class noautolegend;
   title 'Linear Regression';
   reg y=weight x=height;
run;
```

In this example, a single REG statement provides both the scatter plot and the fit line, using a syntax that is similar to the REGRESSIONPLOT statement in the GTL. The NOAUTOLEGEND statement is used with PROC SGPLOT to suppress the display of the automatically generated legend. The results are displayed in Figure 1.6. The two plots are identical.

The REGRESSIONPLOT statement with the DEGREE=3 option finds a cubic polynomial fit function through the points defined by the variables. You can specify DEGREE=2 for a quadratic fit function. The default degree is 1 for ordinary linear regression. The following steps create a cubic fit function:

```
proc template;
   define statgraph classreg;
      begingraph;
         entrytitle 'Cubic Fit Function';
         layout overlay;
            scatterplot y=weight x=height;
            regressionplot y=weight x=height / degree=3;
         endlayout;
      endgraph;
   end;
run;

proc sgrender data=sashelp.class template=classreg;
run;

proc sgplot data=sashelp.class noautolegend;
   title 'Cubic Fit Function';
   reg y=weight x=height / degree=3;
run;
```

The results of these steps are displayed in Figure 1.7 and Figure 1.8. The two plots are identical.

The SG procedures work by writing a template in the GTL and by using it to produce a graph. PROC SGPLOT and PROC SGPANEL have an option on the PROC statement, the TMPLOUT= option, that writes the generated template to a file. You can look at that template and even use or modify it for use with PROC SGRENDER.

Figure 1.9 PROC REG Fit Plot

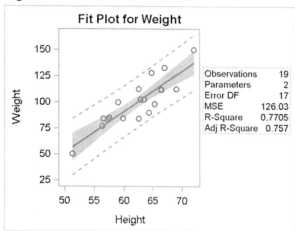

Figure 1.10 PROC GLM Fit Plot

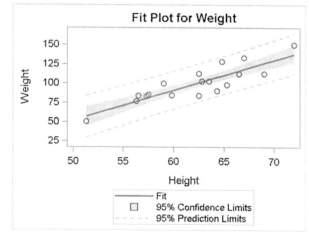

The following step illustrates this option:

```
proc sgplot data=sashelp.class noautolegend tmplout='fittmplt.sas';
   title 'Cubic Fit Function';
   reg y=weight x=height / degree=3;
run;
```

The generated template is displayed in Figure 1.11, and it closely matches the template that is used in this example.

Figure 1.11 Fit Plot Generated Template

```
proc template;
define statgraph sgplot;
begingraph;
EntryTitle "Cubic Fit Function" /;
layout overlay;
   ScatterPlot X=Height Y=Weight / primary=true;
   RegressionPlot X=Height Y=Weight / NAME="REG" LegendLabel="Regression" Degree=3;
endlayout;
endgraph;
end;
run;
```

You can use SAS/STAT procedures to create fit plots when you want more explicit control of the results and when you want more detailed results. Fit plots are created directly by a number of procedures including the REG, GLM, and TRANSREG procedures. The following steps create fit plots directly with PROC REG and with PROC GLM and produce Figure 1.9 and Figure 1.10:

```
ods graphics on;

proc reg data=sashelp.class;
   model weight = height;
run;

proc glm data=sashelp.class;
   model weight = height;
run;
```

Figure 1.12 Loess Fit Plot with Custom Template

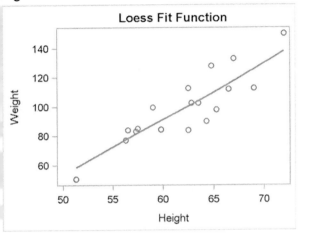

Figure 1.13 Loess Fit Plot from PROC SGPLOT

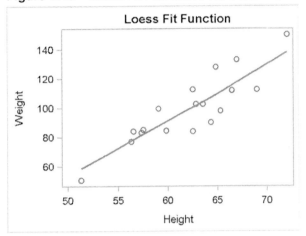

1.2.1 Loess Fit Plot

A loess fit plot consists of a scatter plot of two quantitative variables along with an overlaid nonlinear fit function. The loess fit function is found using a nonparametric locally weighted scatterplot smoothing technique (Cleveland, Devlin, and Grosse 1988). The following steps show how to display a loess fit to a scatter plot:

```
proc template;
   define statgraph classloess;
      begingraph;
         entrytitle 'Loess Fit Function';
         layout overlay;
            scatterplot y=weight x=height;
            loessplot y=weight x=height;
         endlayout;
      endgraph;
   end;
run;

proc sgrender data=sashelp.class template=classloess;
run;

proc sgplot data=sashelp.class noautolegend;
   title 'Loess Fit Function';
   loess y=weight x=height;
run;
```

The LOESSPLOT statement in the GTL and the LOESS statement in PROC SGPLOT fit a nonparametric regression function through the points defined by the variable Weight on the Y axis and the variable Height on the X axis. The NOAUTOLEGEND statement is used with PROC SGPLOT to suppress the display of the automatically generated legend. The results of these steps are displayed in Figure 1.12 and Figure 1.13.

Figure 1.14 Locally Optimal Loess Fit

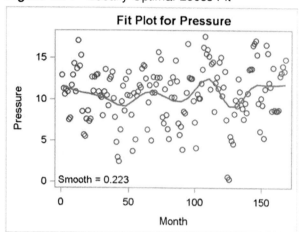

Figure 1.15 Globally Optimal Loess Fit

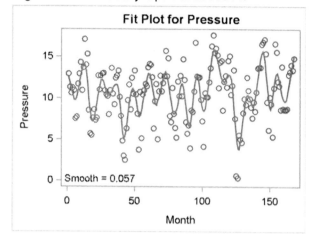

You can use PROC LOESS to create these plots when you want more explicit control of the results and when you are interested in a detailed statistical analysis of your data rather than simply a graph with a fit function.

The following data set contains measurements of monthly averaged atmospheric pressure differences between Easter Island and Darwin, Australia, for a period of 168 months (National Institute of Standards and Technology 1998):

```
data ENSO;
   input Pressure @@;
   Month=_N_;
   format Pressure 4.1;
   format Month 3.0;
   datalines;
12.9  11.3  10.6  11.2  10.9   7.5   7.7  11.7

   ... more lines ...

;
```

You can compute a loess fit in two different ways for these data by using the following statements:

```
ods graphics on;

proc loess data=ENSO;
   model Pressure=Month;
run;

proc loess data=ENSO;
   model Pressure=Month / select=AICC(global);
run;
```

The results of these steps are displayed in Figure 1.14 and Figure 1.15. The plot in Figure 1.14 represents a local optimum, and the plot in Figure 1.15 represents the global optimum. For more information about loess and this example, see "The LOESS Procedure" (*SAS/STAT User's Guide*).

Figure 1.16 Penalized B-Spline with Custom Template **Figure 1.17** Penalized B-Spline Fit from PROC SGPLOT

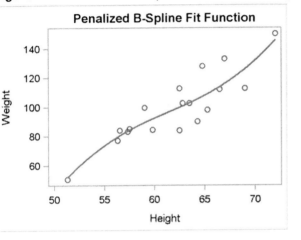

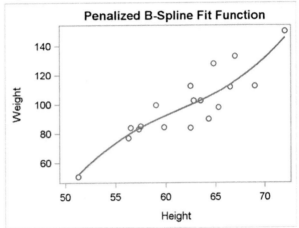

1.2.2 Penalized B-Spline Fit Plot

A penalized B-spline fit plot consists of a scatter plot of two quantitative variables along with an overlaid nonlinear fit function. The penalized B-spline fit function is found using a flexible method (Eilers and Marx 1996) that automatically picks the smoothing parameter that minimizes the corrected AIC criterion, AICC. The following steps show how to make a penalized B-spline fit a plot:

```
proc template;
   define statgraph classpbs;
      begingraph;
         entrytitle 'Penalized B-Spline Fit Function';
         layout overlay;
            scatterplot  y=weight x=height;
            pbsplineplot y=weight x=height;
         endlayout;
      endgraph;
   end;
run;

proc sgrender data=sashelp.class template=classpbs;
run;

proc sgplot data=sashelp.class noautolegend;
   title 'Penalized B-Spline Fit Function';
   pbspline y=weight x=height;
run;
```

The PBSPLINEPLOT statement in the GTL and the PBSPLINE statement in PROC SGPLOT fit a smooth function with an automatically chosen smoothing parameter through the points defined by the variable Weight on the Y axis and the variable Height on the X axis. The results of these steps are displayed in Figure 1.16 and Figure 1.17.

Figure 1.18 Globally Optimal PBSPline Fit

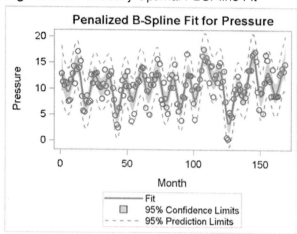

Figure 1.19 Locally Optimal PBSPline Fit

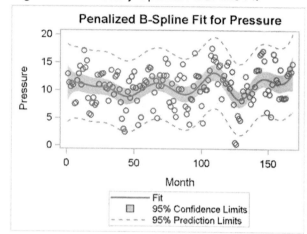

You can use PROC TRANSREG to create these plots when you want more explicit control of the results and when you are interested in a detailed statistical analysis of your data rather than simply a graph with a fit function.

The following data set contains measurements of monthly averaged atmospheric pressure differences between Easter Island and Darwin, Australia, for a period of 168 months (National Institute of Standards and Technology 1998):

```
data ENSO;
   input Pressure @@;
   Month=_N_;
   format Pressure 4.1;
   format Month 3.0;
   datalines;
12.9  11.3  10.6  11.2  10.9   7.5   7.7  11.7

   ... more lines ...

;
```

You can compute a penalized B-spline fit in two different ways for these data by using the following statements:

```
ods graphics on;

proc transreg data=enso;
   model identity(pressure) = pbspline(month);
run;

proc transreg data=enso;
   model identity(pressure) = pbspline(month / sbc lambda=2 10000 range);
run;
```

The results of these steps are displayed in Figure 1.18 and Figure 1.19. The plot in Figure 1.18 represents the global optimum, and the plot in Figure 1.19 represents a local optimum. For more information about penalized B-splines and this example see "The TRANSREG Procedure" (*SAS/STAT User's Guide*).

Figure 1.20   Groups with Custom Template

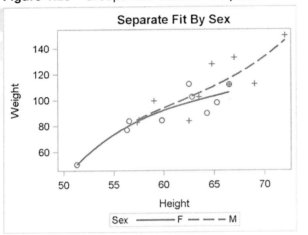

Figure 1.21   Groups with PROC SGPLOT

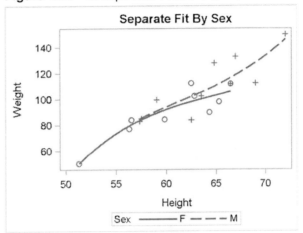

1.2.3   Grouped Regression Fit Plot

Previous examples showed several types of fit plots with a single quantitative dependent variable and a single quantitative independent variable. This example uses a model that additionally has a classification or group variable, and fits a model with separate intercepts and functions for each group. The following steps produce a separate cubic fit function for males and for females:

```
proc sort data=sashelp.class out=class;
   by sex;
run;

proc template;
   define statgraph classgroup;
      begingraph;
         entrytitle 'Separate Fit By Sex';
         layout overlay;
            scatterplot y=weight x=height / group=sex;
            regressionplot y=weight x=height / group=sex degree=3
                                               name='reg';
            discretelegend 'reg' / title='Sex';
         endlayout;
      endgraph;
   end;
run;

proc sgrender data=class template=classgroup;
run;

proc sgplot data=class;
   title 'Separate Fit By Sex';
   reg y=weight x=height / group=sex degree=3;
run;
```

Figure 1.22 Groups with Custom Template

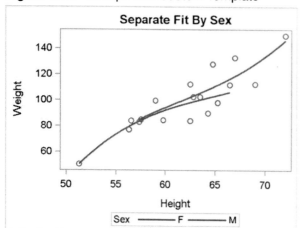

Figure 1.23 Groups with PROC SGPLOT

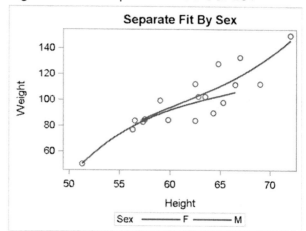

The results are displayed in Figure 1.20 and Figure 1.21. The GROUP= option in both the SCAT-TERPLOT and the REGRESSIONPLOT statement is used to display each of the two groups of observations using the rules specified in the ODS style. With the STATISTICAL style, which is the default style in this book, the females are displayed as blue circles with a blue solid fit function, and the males are displayed as red pluses with a dashed red fit function. Separate computations are performed for each group. For this reason, the input data set must be sorted by the group variable. The REGRESSIONPLOT statement is given a name with the NAME='reg' option. That name is specified in the DISCRETELEGEND statement to produce the legend. A name could have been provided in the SCATTERPLOT statement as well, and it could have been specified in the DISCRETELEGEND statement. However, with this type of plot, the fit functions are the same colors as their corresponding points, so the legend for the REGRESSIONPLOT statement is sufficient. The TITLE= option in the DISCRETELEGEND statement provides a title for the legend. In PROC SGPLOT, the REG statement along with the GROUP= option creates the scatter plot, the fit function, and the legend. The plots are identical.

The following steps make the same plots, but this time by using the **ModStyle** SAS autocall macro to make a color-only style:

```
%modstyle(parent=statistical, name=StatColor)

ods listing style=StatColor;

proc sgrender data=class template=classgroup;
run;

proc sgplot data=class;
   title 'Separate Fit By Sex';
   reg y=weight x=height / group=sex degree=3;
run;
```

The results are displayed in Figure 1.22 and Figure 1.23. The marker for both groups is a circle and the line style is solid. The two graphs are identical. You can also use this macro to specify the color, line pattern, and marker symbol for each group. See the section "An All-Color Style" on page 145 for more on the **ModStyle** macro.[2]

---

[2] The **ModStyle** macro was written by Bob Derr at SAS, and information about it can be found in Kuhfeld (2009).

Figure 1.24 Cubic Fit Plot from PROC TRANSREG

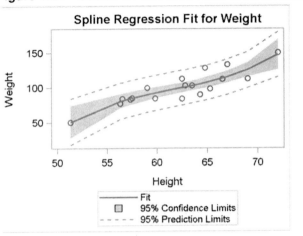

Figure 1.25 Cubic Fit Plot from PROC SGPLOT

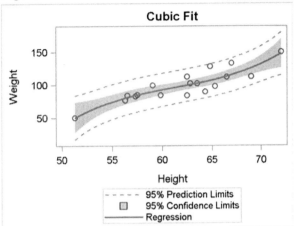

1.2.4 Regression Fit Plot with Confidence and Prediction Limits

This example introduces band plots and series plots. Band plots are used to display confidence limits as a filled band. A series plot is a function that connects a series of points such as predicted values and can be used to display fit functions. The following steps illustrate how much SAS/STAT procedures that create ODS Graphics automatically do for you, how much PROC SGPLOT automatically does for you, and how much you have to do if you want to create the same or similar plots by specifying each of the graph components yourself in either PROC TEMPLATE or PROC SGPLOT. The following steps create a cubic fit plot with confidence and prediction limits:

```
ods graphics on;

proc transreg data=sashelp.class;
   model identity(weight) = spline(height);
   output out=class clm cli p;
run;

proc sgplot data=sashelp.class;
   title 'Cubic Fit';
   reg y=weight x=height / degree=3 cli clm;
run;
```

The first plot is created automatically by PROC TRANSREG and is displayed in Figure 1.24. Only the PROC and MODEL statements are required. The OUTPUT statement and its options create an output data set that is used in subsequent steps, but they are not required to make the fit plot. The second plot is created directly by PROC SGPLOT and is displayed in Figure 1.25. The CLI option (confidence limits for the individual predicted values) in the REG statement produces the outer band, the prediction limits, and the CLM option (confidence limits for mean predicted values) produces the inner band, the confidence limits.

The two plots are similar but not identical. The titles are different, the legend is ordered differently, and the fit function is labeled differently. In addition, PROC SGPLOT automatically creates smoother functions by computing predicted values, confidence, and prediction limits for interpolated values. PROC TRANSREG, by default, only computes these values for each observed X value. This difference is typically only noticeable for small data sets such as the one used in this example.

The following steps use the output data set from PROC TRANSREG and produce the same plots by manually requesting the scatter plot, the fit function, the prediction limits, and the confidence limits, each with a separate statement:

```
proc template;
   define statgraph classfit;
      begingraph;
         entrytitle 'Cubic Fit';
         layout overlay;
            bandplot limitupper=cmuweight limitlower=cmlweight x=height /
                  connectorder=axis outlineattrs=GraphConfidence
                  name='95% Confidence Limits';
            bandplot limitupper=ciuweight limitlower=cilweight x=height /
                  connectorder=axis display=(outline)
                  outlineattrs=GraphPredictionLimits
                  name='95% Prediction Limits';
            scatterplot y=weight x=height;
            seriesplot y=pweight x=height / name='Fit' legendlabel='Fit'
                        connectorder=xaxis lineattrs=GraphFit;
            discretelegend 'Fit' '95% Confidence Limits'
                        '95% Prediction Limits';
         endlayout;
      endgraph;
   end;
run;

proc sgrender data=class template=classfit;
run;

proc sort data=class out=sorted;
   by height;
run;

proc sgplot data=sorted;
   title 'Cubic Fit';
   band upper=cmuweight lower=cmlweight x=height /
         fillattrs=GraphConfidence
         name='b1' legendlabel='95% Confidence Limits';
   band upper=ciuweight lower=cilweight x=height /
         nofill lineattrs=GraphPredictionLimits
         name='b2' legendlabel='95% Prediction Limits';
   scatter y=weight x=height;
   series y=pweight x=height / legendlabel='Fit' lineattrs=GraphFit name='f';
   keylegend 'f' 'b1' 'b2';
run;
```

The following GTL statement produces the confidence limits:

```
bandplot limitupper=cmuweight limitlower=cmlweight x=height /
     connectorder=axis outlineattrs=GraphConfidence
     name='95% Confidence Limits';
```

A filled band is displayed across the range of the X axis variable Height between the lower limit variable cmlweight and the upper limit variable cmuweight. The CONNECTORDER=AXIS option connects the data points left to right along the X axis. The OUTLINEATTRS= option specifies the GraphConfidence style element for the outline. The NAME= option provides a statement name so that the confidence limit information can be added to the legend.

The following GTL statement produces the prediction limits:

```
bandplot limitupper=ciuweight limitlower=cilweight x=height /
     connectorder=axis display=(outline)
     outlineattrs=GraphPredictionLimits
     name='95% Prediction Limits';
```

The limits are displayed across the range of the X axis variable Height between the lower limit variable cilweight and the upper limit variable ciuweight. The CONNECTORDER=AXIS option connects the data points left to right along the X axis. The DISPLAY= option is used to display only the outline of the band rather than the entire band. The OUTLINEATTRS= option specifies the GraphPredictionLimits style element for the outline. The NAME= option provides a statement name so that the prediction limit information can be added to the legend.

The following GTL statement produces the scatter plot:

```
scatterplot y=weight x=height;
```

It plots the variable Weight on the Y axis and the variable Height on the X axis.

The following GTL statement produces the series plot with the fit function:

```
seriesplot y=pweight x=height / name='Fit' legendlabel='Fit'
              connectorder=xaxis lineattrs=GraphFit;
```

The predicted values were previously computed by PROC TRANSREG, so the SERIESPLOT statement only connects them to provide the fit function. The label for the legend is 'Fit'. The LINEATTRS= specifies the GraphFit style for the fit line. Like the BANDPLOT statements, the CONNECTORDER=AXIS and NAME= options are specified.

The following GTL statement produces the legend from the three named statements:

```
discretelegend 'Fit' '95% Confidence Limits' '95% Prediction Limits';
```

The results are displayed in Figure 1.26.

Figure 1.26 Cubic Fit Plot with Custom Template

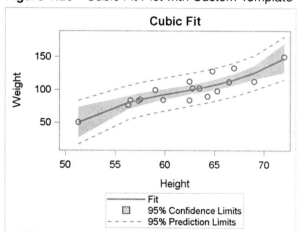

Figure 1.27 Manually Constructed with PROC SGP

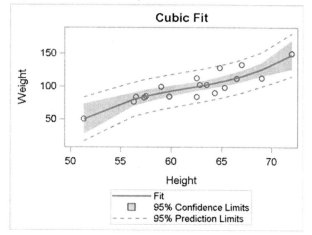

The BAND statements in PROC SGPLOT require that the data set be sorted by the X axis variable, Height. Hence the first step for the next plot is PROC SORT. The first plotting statement in PROC SGPLOT is the following:

```
band upper=cmuweight lower=cmlweight x=height / fillattrs=GraphConfidence
    name='b1' legendlabel='95% Confidence Limits';
```

A filled band is displayed across the range of the X axis variable Height between the lower limit variable cmlweight and the upper limit variable cmuweight. The FILLATTRS= option specifies the **GraphConfidence** style element for the fill (the band). The NAME= option provides a statement name so that the confidence limit information can be added to the legend. The LEGENDLABEL= option provides the text that is displayed in the legend along with the confidence limits information.

The following PROC SGPLOT statement produces the prediction limits:

```
band upper=ciuweight lower=cilweight x=height /
    nofill lineattrs=GraphPredictionLimits
    name='b2' legendlabel='95% Prediction Limits';
```

The limits are displayed across the range of the X axis variable Height between the lower limit variable cilweight and the upper limit variable ciuweight. The NOFILL option produces just an outline instead of a band. The LINEATTRS= option specifies the **GraphPredictionLimits** style element for the outline. The NAME= option provides a statement name so that the prediction limit information can be added to the legend. The LEGENDLABEL= option provides the text that is displayed in the legend along with the prediction limits information.

The following PROC SGPLOT statement produces the scatter plot:

```
scatter y=weight x=height;
```

It plots the variable Weight on the Y axis and the variable Height on the X axis.

The following PROC SGPLOT statement produces the series plot with the fit function:

```
series y=pweight x=height / legendlabel='Fit' lineattrs=GraphFit name='f';
```

Figure 1.28 Prediction Limits Outlined

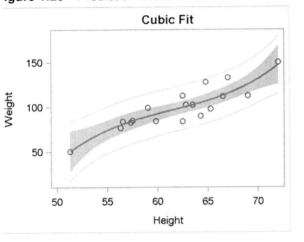

Figure 1.29 Confidence Style Elements

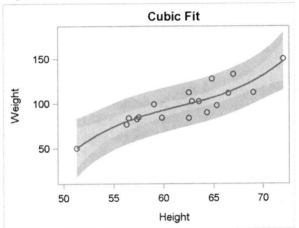

The predicted values were previously computed by PROC TRANSREG, so it only connects them to provide the fit function. The label for the legend is 'Fit'. The LINEATTRS= specifies the `GraphFit` style for the fit line. Like the BAND statements, the NAME= option is specified.

The following PROC SGPLOT statement produces the legend from the three named statements:

```
keylegend 'f' 'b1' 'b2';
```

The results are displayed in Figure 1.27.

Both the REG statement in PROC SGPLOT and the REGPLOT statement in the GTL accept the following options and many others:

- ALPHA=0.05, the default, produces 95% confidence and prediction limits. Specify ALPHA=0.01 for 99% confidence and prediction limits.

- CURVELABEL=*'string'* specifies a label for the regression function.

The previous steps used the GTL, PROC TEMPLATE, and PROC SGRENDER to display regression results that were computed by another procedure (PROC TRANSREG). This next series of steps does all of the computations within GTL, PROC TEMPLATE, and PROC SGRENDER and displays the results. The following examples illustrate the MODELBAND statement and also the technique of using and overriding style elements to control the appearance of the confidence and prediction limits.

The following steps create Figure 1.28:

```
proc template;
   define statgraph classfit1;
      begingraph;
         entrytitle 'Cubic Fit';
         layout overlay;
            modelband "cliband" / display=(outline);
            modelband "clmband";
            scatterplot y=weight x=height;
            regressionplot y=weight x=height / degree=3
                           clm="clmband" cli="cliband";
         endlayout;
      endgraph;
   end;
run;

proc sgrender data=sashelp.class template=classfit1;
run;
```

The REGRESSIONPLOT statement with the CLM="clmband" option displays the confidence limits according to the options specified on the MODELBAND statement named "clmband". This MODELBAND statement has no options so a default band plot is produced, which consists of a filled band with no outline. Similarly, the CLM="cliband" option displays the prediction limits according to the options specified on the MODELBAND statement named "cliband". This MODELBAND statement specifies that an outline of the band is to be displayed, but the band is not filled. Statements are executed in the order that they are specified in the template. The prediction limits are drawn first, then the confidence limits, then the scatter plot, then the fit function. Plot elements that are placed later can obscure previously placed plot elements. Hence, you do not want to draw the band plots last, since you will not be able to see the fit function and many of the points.

The following steps create Figure 1.29:

```
proc template;
   define statgraph classfit2;
      begingraph;
         entrytitle 'Cubic Fit';
         layout overlay;
            modelband "cliband" / fillattrs=GraphConfidence;
            modelband "clmband" / fillattrs=GraphConfidence2;
            scatterplot y=weight x=height;
            regressionplot y=weight x=height / degree=3
                           clm="clmband" cli="cliband";
         endlayout;
      endgraph;
   end;
run;

proc sgrender data=sashelp.class template=classfit2;
run;
```

In this example, the FillAttrs=GraphConfidence option uses the GraphConfidence style element to control the appearance of the prediction limits and the FillAttrs=GraphConfidence2 option uses the GraphConfidence2 style element to control the appearance of the confidence limits.

Figure 1.30 Outline, Fill, Transparency

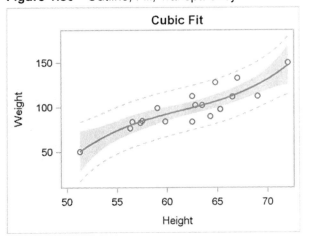

Figure 1.31 Style Element Overrides

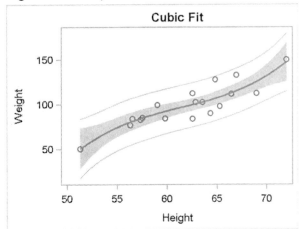

The following steps create Figure 1.30:

```
proc template;
   define statgraph classfit3;
      begingraph;
         entrytitle 'Cubic Fit';
         layout overlay;
            modelband "cliband" / outlineattrs=GraphPredictionLimits
                                   display=(outline)
                                   datatransparency=0.5;
            modelband "clmband" / fillattrs=GraphConfidence
                                  datatransparency=0.5;
            scatterplot y=weight x=height;
            regressionplot y=weight x=height / degree=3
                           clm="clmband" cli="cliband";
         endlayout;
      endgraph;
   end;
run;

proc sgrender data=sashelp.class template=classfit3;
run;
```

Figure 1.30 has a style that is similar to the graph produced directly by PROC TRANSREG (shown in Figure 1.24) and to the graph produced directly by PROC SGPLOT (shown in Figure 1.25). The prediction limits are displayed as an outline without fill due to the DISPLAY=(OUTLINE) option. The style of the line is controlled by the **OutLineAttrs=GraphPredictionLimits** option, which uses the **GraphPredictionLimits** style element, which specifies a dashed line in this style. The DATATRANSPARENCY=0.5 option is specified so that the bands and outlines are 50% as transparent as the default. The confidence limits are displayed as a band without an outline due to the default setting of the DISPLAY= option, DISPLAY=(FILL). The style of the fill is controlled by the **FillAttrs=GraphConfidence** option, which uses the **GraphConfidence** style element.

The following steps create Figure 1.31:

```
proc template;
   define statgraph classfit4;
      begingraph;
         entrytitle 'Cubic Fit';
         layout overlay;
            modelband "cliband" /
                        outlineattrs=GraphPredictionLimits(pattern=solid)
                        display=(outline)
                        datatransparency=0.5;
            modelband "clmband" /
                        fillattrs=GraphConfidence(color=cx88AAAA)
                        datatransparency=0.5;
            scatterplot y=weight x=height;
            regressionplot y=weight x=height / degree=3
                        clm="clmband" cli="cliband";
         endlayout;
      endgraph;
   end;
run;

proc sgrender data=sashelp.class template=classfit4;
run;
```

These steps are similar to the preceding steps. However, one part of the `GraphPredictionLimits` style element is overridden, namely the line pattern. The option `OutLineAttrs=Graph-PredictionLimits(Pattern=Solid)` creates prediction limits using all style elements from the `GraphPredictionLimits` element except for the default line style. Similarly, the `FillAttrs=GraphConfidence(Color=cx88AAAA)` option overrides the color of the `GraphConfidence` style element. All colors can be specified in values of the form CX*rrggbb*, where the last six characters specify RGB (red, green, blue) values on the hexadecimal scale of 00 to FF (or 0 to 255 base 10). HLS (hue/light/saturation) color specifications and all other color specifications available in SAS/GRAPH are also available in ODS Graphics.

Figure 1.32 Inset Text with the GTL

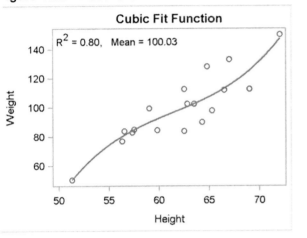

Figure 1.33 Special Characters with the GTL

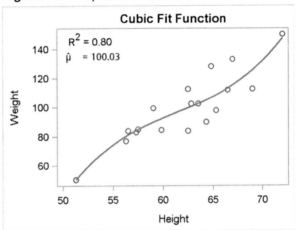

Figure 1.34 Special Characters with PROC SGPLOT

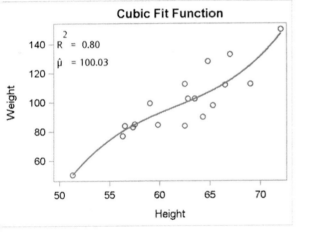

Figure 1.35 Alternative Syntax with the GTL

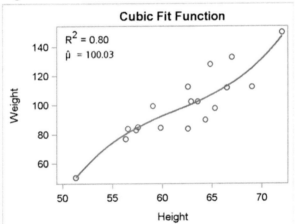

1.2.5 Text Insets and Special Characters

This section shows how to insert text into a graph including text with superscripts (R^2) and a Greek letter with a special character ($\hat{\mu}$). The graph is the cubic fit function from the section "Regression Fit Plot" on page 6. The following step runs PROC TRANSREG with a cubic polynomial model and creates an ODS output data set with regression fit statistics including R square and the mean of the dependent variable:

```
proc transreg data=sashelp.class ss2;
   ods output fitstatistics=fs;
   model identity(weight) = spline(height);
run;
```

The following step stores the R square in a macro variable, R2, and the mean in a macro variable, Mean:

```
data _null_;
   set fs;
   if _n_ = 1 then call symputx('R2', put(value2, 4.2), 'G');
   if _n_ = 2 then call symputx('mean', put(value1, best6.), 'G');
run;
```

The R square value is stored with a format of 4.2, and the mean is stored with the BEST6. format. Both are stored in the global macro symbol table. These values are inserted into plots as inset text entries in the next steps.

The following steps create a plot with the R square and the mean in the top left corner:

```
proc template;
   define statgraph classreg1;
      mvar r2 mean;
      begingraph;
         entrytitle 'Cubic Fit Function';
         layout overlay;
            entry halign=left 'R' {sup '2'} ' = ' r2
                  ",   Mean = " mean / valign=top;
            scatterplot y=weight x=height;
            regressionplot y=weight x=height / degree=3;
         endlayout;
      endgraph;
   end;
run;


proc sgrender data=sashelp.class template=classreg1;
run;
```

The MVAR statement specifies the two macro variables, R2 and Mean, so that they can be used in other parts of the template. The ENTRY statement in this example adds a single line of text to the top left of the plot consisting of an "R^2", an equal sign, the formatted R square value, a comma and spaces, "Mean =", and the mean value. The inset text entry is created from a series of values on the ENTRY statement including constant text, special characters, and values retrieved from macro variables. The syntax `'R' {sup '2'}` shows the syntax for superscripts. (Similarly, the syntax for subscripts is as follows: `'x' {sub 'i'}`.) Notice that the macro variables are specified without ampersands. Therefore, their values are not retrieved just before PROC TEMPLATE compiles the template; rather, the values are retrieved at the time that PROC SGRENDER is run. This approach enables you to create the template once and use it repeatedly for different analyses with different values for the two statistics. Also notice that the specification `{sup '2'}` does not appear in quotes. If it had appeared in quotes, then "{sup '2'}" would appear in the graph and not "^{2}". The results are displayed in Figure 1.32.

The following steps insert the same information into the plot, but with two differences:

```
proc template;
   define statgraph classreg2;
      mvar r2 mean;
      begingraph;
         entrytitle 'Cubic Fit Function';
         layout overlay;
            layout gridded / autoalign=(topright topleft
                                        bottomright bottomleft);
               entry 'R' {sup '2'} ' = ' r2;
               entry "   (*ESC*){unicode mu}(*ESC*){unicode hat}    = " mean /
                     textattrs=GraphValueText(family=GraphUnicodeText:FontFamily);
               endlayout;
            scatterplot y=weight x=height;
            regressionplot y=weight x=height / degree=3;
         endlayout;
      endgraph;
   end;
run;

proc sgrender data=sashelp.class template=classreg2;
run;
```

The results of this step are displayed in Figure 1.33. In this example, the text is split onto two lines, and the mean is labeled by "$\hat{\mu}$" instead of "Mean". The LAYOUT GRIDDED statement groups the two ENTRY statements, which enables them to be moved and positioned as a group with one immediately following the other. You can place any number of ENTRY statements between the LAYOUT and the ENDLAYOUT statement. The LAYOUT GRIDDED statement has a number of options. For example, you can specify the options OPAQUE=TRUE, BORDER=TRUE, and BACKGROUNDCOLOR=*style-reference* | *color* to create an opaque box with a background and color. In this example, the AUTOALIGN= option is used to specify a list of preferred positions. ODS Graphics places the entries in the first position in the list available position. In this example, there are data in the top right, so the entries are placed in the top left.

The second ENTRY statement specifies "$\hat{\mu}$" with the specification `(*ESC*){UNICODE mu}(*ESC*){UNICODE hat}` which appears in quotes. Special characters are specified as Unicode characters with an escape sequence. The `(*ESC*)` specification is a flag that indicates that a special character follows. When you use the `(*ESC*)` specification is discussed later on in this section. For now, simply note that the `{sup '2'}` specification is not in quotes and is not escaped, whereas the Unicode specifications are in quotes and are escaped. The TEXTATTRS= option is specified so that a font that has the Unicode characters is used. Notice that the font is different for the two lines. Also notice that extra white space was added to slightly better align the parts of each line.

Greek letters are specified by: `{UNICODE letter}`. Lowercase Greek letters are specified by name (for example, **ALPHA**, **BETA**, **GAMMA**, ..., **OMEGA**). Uppercase Greek letters are specified by appending an **_U** to the name (for example, **ALPHA_U**, **BETA_U**, **GAMMA_U**, ..., **OMEGA_U**). Additional keywords include **BAR**, **BAR2**, **HAT**, **TILDE**, and **PRIME**. These Unicode values are part of the Unicode range called combining (nonspacing) diacritical marks. This means that they are drawn superimposed on the previous character rather than in the next position. These characters should always immediately follow the character to be over struck. Hence, `{UNICODE MU}{UNICODE HAT}` (which is correct), is different from `{UNICODE HAT}{UNICODE MU}` (which is backwards and incorrect), which is different from `{UNICODE MU} {UNICODE HAT}` (which includes a space and so is also incorrect).

The Unicode Consortium (`http://unicode.org/`) provides a page of character codes at `http://www.unicode.org/charts/charindex.html`.

The following step uses PROC SGPLOT and produces the plot displayed in Figure 1.34, which is similar to the plot displayed in Figure 1.33, which was created in the previous steps by PROC TEMPLATE and PROC SGRENDER:

```
proc sgplot data=sashelp.class noautolegend;
   title 'Cubic Fit Function';
   inset "R(*ESC*){sup '2'} =  &r2"
         "(*ESC*){unicode mu}(*ESC*){unicode hat}   = &mean" /
         position=topleft;
   reg y=weight x=height / degree=3;
run;
```

The INSET statement is used in this example to produce two entries since two quoted strings are specified. In the INSET statement (unlike the ENTRY statement in the GTL), each string forms a new line, and each string must be fully quoted, even when special characters are provided. Each Unicode, superscript, and subscript specification must be escaped since each must appear inside a quoted string. This is how ODS Graphics knows that a special character is coming. Otherwise, each of these specifications appears "as is" in the graph. Since PROC SGPLOT does not have an MVAR statement like the GTL, the macro variables appear inside of the double quotes and appear with an ampersand. Their values are resolved at the time that PROC SGPLOT is run. Note that the specification `"R(*ESC*){sup '2'} = &r2"` contains outer double quotes and inner single quotes. Double quotes are required for the outer quotes so that the macro variables will resolve. Double quotes could be used for the inner quotes as well, but they have to be doubled so that they do not end the string (for example, as follows: `"R(*ESC*){sup ""2""} = &r2"`). No TEXTATTRS= option is needed in PROC SGPLOT. It automatically switches to the appropriate font when the Unicode characters are encountered.

The final steps in this example use PROC TEMPLATE and PROC SGRENDER and produce the plot without any escape characters. The following steps produce Figure 1.35:

```
proc template;
   define statgraph classreg3;
      mvar r2 mean;
      begingraph;
         entrytitle 'Cubic Fit Function';
         layout overlay;
            layout gridded / autoalign=(topright topleft
                                        bottomright bottomleft);
               entry 'R' {sup '2'} ' = ' r2;
               entry '  ' {unicode mu}{unicode hat} '  = ' mean /
                     textattrs=GraphValueText(family=GraphUnicodeText:FontFamily
               endlayout;
            scatterplot y=weight x=height;
            regressionplot y=weight x=height / degree=3;
         endlayout;
      endgraph;
   end;
run;

proc sgrender data=sashelp.class template=classreg3;
run;
```

Figure 1.36 Bar Chart with Custom Template

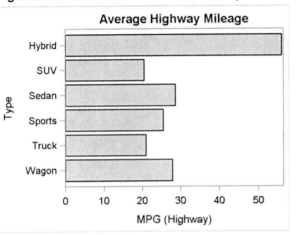

Figure 1.37 Bar Chart with PROC SGPLOT

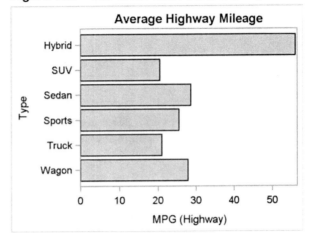

Figure 1.38 Bar Chart with Custom Template

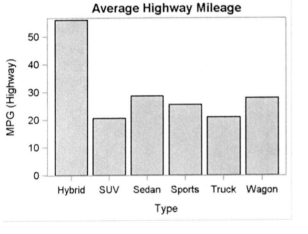

Figure 1.39 Bar Chart with PROC SGPLOT

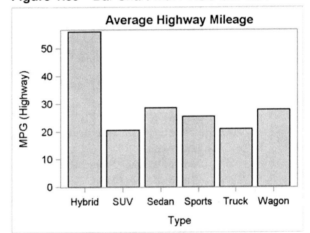

1.3 Bar Chart

A bar chart is a graph with rectangular bars whose lengths are proportional to the values that they represent (for example, means, percents, sums, and so on). Bar charts are used to compare the values of a quantitative variable at two or more levels of a categorical variable. The following steps produce a horizontal bar chart of mileage by vehicle type:

```
proc template;
   define statgraph barchart;
      begingraph;
         entrytitle 'Average Highway Mileage';
         layout overlay;
            barchart x=type y=mpg_highway / stat=mean orient=horizontal;
         endlayout;
      endgraph;
   end;
run;
```

```
proc sort data=sashelp.cars out=cars;
   by descending type;
run;

proc sgrender data=cars template=barchart;
run;

proc sgplot data=sashelp.cars;
   title 'Average Highway Mileage';
   hbar type / response=mpg_highway stat=mean;
run;
```

In the GTL, the BARCHART statement along with the ORIENT=HORIZONTAL option create a horizontal bar chart. With the STAT=MEAN option, the mean of the Y= variable is displayed for each category of the X= discrete variable. The options for STAT= include: FREQ, PCT, SUM, and MEAN. The data are sorted before running PROC SGRENDER so that the vehicle types will appear in the same sorted order that PROC SGPLOT uses. Without this step, the vehicle types appear in the order that they appear in the original data set. PROC SORT needs to sort the data in descending order, which will display the bars in alphabetical order reading from the top to the bottom. PROC SGPLOT uses an HBAR statement, specifies the discrete variable first in front of the slash, and specifies the response variable in the RESPONSE= option. The options for STAT= include: FREQ, SUM, and MEAN. The results are displayed in Figure 1.36 and Figure 1.37.

The following steps produce a vertical bar chart of mileage by vehicle type:

```
proc template;
   define statgraph barchart;
      begingraph;
         entrytitle 'Average Highway Mileage';
         layout overlay;
            barchart x=type y=mpg_highway / stat=mean;
         endlayout;
      endgraph;
   end;
run;

proc sort data=sashelp.cars out=cars;
   by type;
run;

proc sgrender data=cars template=barchart;
run;

proc sgplot data=sashelp.cars;
   title 'Average Highway Mileage';
   vbar type / response=mpg_highway stat=mean;
run;
```

PROC SORT is used to sort the data in ascending order, which will display the bars in alphabetical order reading left to right. The other options here are an obvious variation on the options in the previous steps. In the GTL the default orientation is ORIENT=VERTICAL, and in PROC SGPLOT the VBAR statement produces a vertical bar chart. The results are displayed in Figure 1.38 and Figure 1.39.

Figure 1.40 Parameterized Vertical Bar Chart

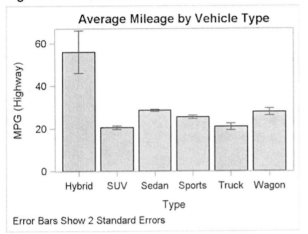

Figure 1.41 Parameterized Horizontal Bar Chart

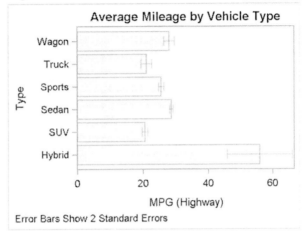

The following step adds error bars showing two standard errors to the vertical bar chart by using the NUMSTD=2 and LIMITSTAT=STDERR options:

```
proc sgplot data=sashelp.cars noautolegend;
   title 'Average Mileage by Vehicle Type';
   vbar type / response=mpg_highway stat=mean numstd=2 limitstat=stderr;
   footnote justify=left 'Error Bars Show 2 Standard Errors';
run;
```

The results of this step are not displayed, but they are almost identical to the graph displayed in Figure 1.40.

The preceding bar chart examples provide ODS Graphics with raw data and with multiple observations in each group. ODS Graphics through either the BARCHART statement in PROC SGRENDER or the VBAR or HBAR statement in PROC SGPLOT does the computations to get the means, percents, or other statistics and produce the results. The remainder of this section uses only PROC TEMPLATE and PROC SGRENDER with the BARCHARTPARM statement. Statements with a name that contains "PARM" do not do computations to summarize the data. You provide a data set with one observation per group and the information that needs to be displayed. The advantage of using the BARCHARTPARM statement is you can have precise control over how the groups are created. You can create this data set from the raw data using a procedure such as PROC SUMMARY or PROC MEANS. The following steps use the BARCHARTPARM statement to produce a bar chart showing group means with error bars:

```
proc summary data=sashelp.cars nway;
   class type;
   var mpg_highway;
   output out=mileage mean=mean stderr=stderr;
run;
```

```
proc template;
   define statgraph barchartparm;
      begingraph;
         entrytitle    "Average Mileage by Vehicle Type";
         entryfootnote halign=left textattrs=GraphValueText
                       "Error Bars Show 2 Standard Errors";
         layout overlay;
            barchartparm x=type y=mean /
                    errorlower=eval(mean - 2 * stderr)
                    errorupper=eval(mean + 2 * stderr);
         endlayout;
      endgraph;
   end;
run;


proc sgrender data=mileage template=barchartparm;
run;
```

The results are displayed in Figure 1.40. In the GTL, the ENTRYTITLE statement provides the title and the ENTRYFOOTNOTE statement provides the footnote. Options are specified on the ENTRYFOOTNOTE statement to control the position and the appearance of the footnote. The option HALIGN=LEFT aligns the footnote on the left; the default is centered. The TEXTATTRS= option uses the `GraphValueText` style element to display the footnote. The `GraphFootnoteText` style element is the default, which in the STATISTICAL style uses an italic font.

The BARCHARTPARM statement specifies the variable Type on the X= option and the variable Mean on the Y= option. These variables are computed in the PROC SUMMARY step that follows the PROC TEMPLATE step. The GTL can perform computations and use variables that are functions of the input data. These computations are always performed in the context of the EVAL function. This BARCHARTPARM statement creates lower and upper error bars using the syntax: `ErrorLower=eval(Mean - 2 * StdErr) ErrorUpper=eval(Mean + 2 * StdErr)`. The variables Mean and StdErr are computed by PROC SUMMARY. The expressions in the EVAL function add and subtract two standard errors from the mean to make the error bars.

PROC SUMMARY creates an output data set with descriptive statistics. The NWAY option outputs only n-way (in this case one-way) frequencies. This suppresses the creation of an observation with the overall mean. One observation is created for each level of the CLASS statement variable Type with the group means computed from the variable MPG_Highway. The means are stored in the variable Mean, and the standard errors are stored in the variable StdErr. PROC SGRENDER produces Figure 1.40.

The following steps use the ORIENT=HORIZONTAL option and create the horizontal bar chart that is displayed in Figure 1.41:

```
proc template;
   define statgraph barchartparm;
      begingraph;
         entrytitle    "Average Mileage by Vehicle Type";
         entryfootnote halign=left textattrs=GraphValueText
                       "Error Bars Show 2 Standard Errors";
         layout overlay;
            barchartparm x=type y=mean / orient=horizontal
                     datatransparency=0.75
                     errorlower=eval(mean - 2 * stderr)
                     errorupper=eval(mean + 2 * stderr);
         endlayout;
      endgraph;
   end;
run;


proc sgrender data=mileage template=barchartparm;
run;
```

In this step, the DATATRANSPARENCY=0.75 option is specified so that the bars are 75% as transparent as the default. Values range from 0 to 1 with values closer to 1 creating bars that are nearly invisible. The default is DATATRANSPARENCY=0.

Figure 1.42   Histogram with Custom Template

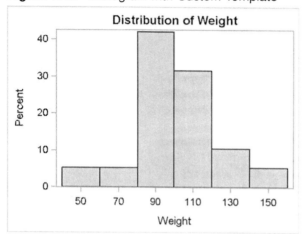

Figure 1.43   Histogram with PROC SGPLOT

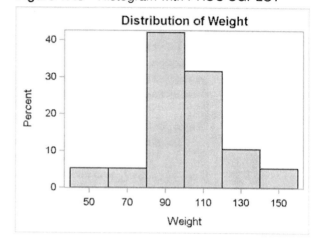

Figure 1.44   Histogram with Standard Axis

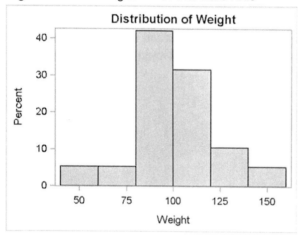

Figure 1.45   Histogram with Bin End Ticks

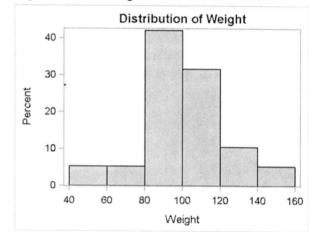

1.4   Histogram

A histogram displays tabulated frequencies, percents, or proportions as bars. Each bar represents a nonoverlapping interval of a quantitative variable. The following steps produce a histogram and show the distribution of weight in the Sashelp.Class data set:

```
proc template;
    define statgraph classhist;
        begingraph;
            entrytitle 'Distribution of Weight';
            layout overlay;
                histogram weight;
            endlayout;
        endgraph;
    end;
run;
```

```
proc sgrender data=sashelp.class template=classhist;
run;

proc sgplot data=sashelp.class;
   title 'Distribution of Weight';
   histogram weight / showbins;
run;
```

Both the GTL and the PROC SGPLOT HISTOGRAM statement name the variable to display. In PROC SGPLOT, the SHOWBINS option specifies that the midpoints of the value bins are used to create the tick marks for the horizontal axis. By default, the tick marks are created at regular intervals based on the minimum and maximum values.

The results are displayed in Figure 1.42 and Figure 1.43. The two graphs are identical.

The following steps create a histogram with axes that are not binned:

```
proc template;
   define statgraph classhist;
      begingraph;
         entrytitle 'Distribution of Weight';
         layout overlay;
            histogram weight / binaxis=false;
         endlayout;
      endgraph;
   end;
run;

proc sgrender data=sashelp.class template=classhist;
run;

proc sgplot data=sashelp.class;
   title 'Distribution of Weight';
   histogram weight;
run;
```

Both graphs are identical, and one is displayed in Figure 1.44. Ticks range from a minimum to a maximum based on an increment, all of which are independent of the bin width.

Figure 1.46 Histogram with Bin Width of 20

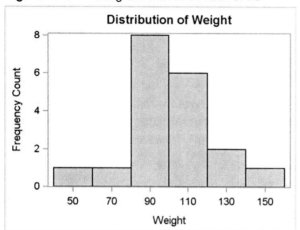

Figure 1.47 Histogram with Bin Width of 25

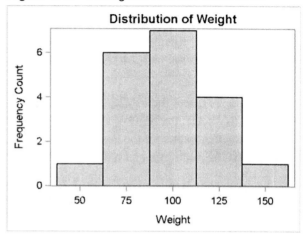

The following step produces a histogram with bin labels at the end of each bin:

```
proc template;
   define statgraph classhist;
      begingraph;
         entrytitle 'Distribution of Weight';
         layout overlay;
            histogram weight / endlabels=true;
         endlayout;
      endgraph;
   end;
run;

proc sgrender data=sashelp.class template=classhist;
run;
```

The results are displayed in Figure 1.45.

The preceding histogram examples provide ODS Graphics with raw data and with multiple observations in each data range. ODS Graphics does the computations and produces the results. The remainder of this section uses only PROC TEMPLATE and PROC SGRENDER with the HISTOGRAMPARM statement. This statement requires the data to be summarized. You provide a data set with one observation per bar and the information that needs to be displayed. The advantage of using the HISTOGRAMPARM statement is you can have precise control over how the data are grouped. You can create this data set from the raw data using a procedure such as PROC FREQ. The following steps use the HISTOGRAMPARM statement to produce a histogram:

```
data class;
   set sashelp.class;
   w1 = round(weight - 10, 20) + 10;
   w2 = round(weight, 25);
   label w1 = 'Weight' w2 = 'Weight';
run;
```

```
proc freq data=class;
    tables w1 / out=freqs1 noprint;
    tables w2 / out=freqs2 noprint;
run;

proc template;
    define statgraph classhist;
        mvar x;
        begingraph;
            entrytitle 'Distribution of Weight';
            layout overlay;
                histogramparm x=x y=count;
            endlayout;
        endgraph;
    end;
run;

%let x = w1;
proc sgrender data=freqs1 template=classhist;
run;

%let x = w2;
proc sgrender data=freqs2 template=classhist;
run;
```

The first DATA step creates two new variables, W1 and W2. The variable W1 contains each individual's weight, rounded, with increments between the rounded values of 20. Each rounded value is a multiple of 10 (50, 70, 90 ...), but not a multiple of 20 (40, 60, 70 ...). This is accomplished by subtracting 10, rounding to the nearest multiple of 20, then adding 10 to the rounded values. The variable W1 contains each individual's weight, rounded to the nearest multiple of 25. PROC FREQ is used to tabulate the number of values in each weight group for both of the variables. The results are stored in SAS data sets in the variable Count.

The HISTOGRAMPARM statement specifies the frequency variable for each interval on the Y= option. The X= option specifies the variable that contains the midpoints of the intervals. In this case, rather than naming a variable in the data set, the specification X=X specifies the macro variable X. The MVAR statement specifies that the value of the macro variable X is retrieved at the time that PROC SGRENDER is run (or more generally at the time that the procedure that uses the template is run, not the time when PROC TEMPLATE is run). Notice that the macro variables are specified without ampersands. Therefore, their values are not substituted at the time that SAS ordinarily processes macro variables and performs substitutions. The value of the macro variable X is W1 at the time that the first PROC SGRENDER step is run and so the X= specification becomes in effect X=W1. For the second PROC SGRENDER step, the X= specification becomes in effect X=W2. The MVAR statement is used instead of the direct specification so that the same template can be used to make both plots, one for W1 and the other for W2.

Figure 1.48 Density with Custom Template

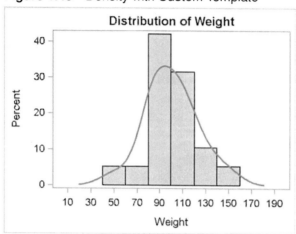

Figure 1.49 Density with PROC SGPLOT

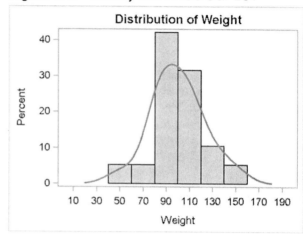

1.5 Density Plot

A probability density function describes the density of probability at each point in the sample space. The probability of a random variable falling within a given set is given by the integral of the density over the set. Common probability density functions include the normal, the Chi square, the F, and the Student t. Kernel density estimation is a nonparametric way of estimating the probability density function of a random variable. The following steps produce a histogram and a kernel density function and show the distribution of weight in the Sashelp.Class data set:

```
proc template;
   define statgraph classhistden;
      begingraph;
         entrytitle 'Distribution of Weight';
         layout overlay;
            histogram weight;
            densityplot weight / kernel();
         endlayout;
      endgraph;
   end;
run;

proc sgrender data=sashelp.class template=classhistden;
run;

proc sgplot data=sashelp.class noautolegend;
   title 'Distribution of Weight';
   histogram weight / showbins;
   density weight / type=kernel;
run;
```

Both procedures have a HISTOGRAM statement like before. In the GTL, the DENSITYPLOT statement uses the specification KERNEL() to create the kernel density. In PROC SGPLOT, the DENSITY statement uses the specification TYPE=KERNEL to create the kernel density. The results are displayed in Figure 1.48 and Figure 1.49.

The following steps produce a histogram and a normal density function and show the distribution of weight in the Sashelp.Class data set:

```
proc template;
   define statgraph classhistden2;
      begingraph;
         entrytitle 'Distribution of Weight';
         layout overlay;
            histogram weight;
            densityplot weight / normal();
         endlayout;
      endgraph;
   end;
run;

proc sgrender data=sashelp.class template=classhistden2;
run;

proc sgplot data=sashelp.class noautolegend;
   title 'Distribution of Weight';
   histogram weight / showbins;
   density weight / type=normal;
run;
```

The results of these steps differ from the previous steps only in the shape of the density function and are not shown.

You can use PROC UNIVARIATE or PROC KDE to directly create histograms and perform kernel density estimation when you want more explicit control of the results and when you are interested in a detailed statistical analysis of your data rather than simply a graph.

The following steps perform univariate and bivariate kernel density estimation with PROC KDE and PROC UNIVARIATE:

```
ods graphics on;

proc kde data=sashelp.class;
   univar weight;
run;

proc kde data=sashelp.class;
   bivar height weight;
run;
```

```
proc univariate data=sashelp.class;
   var weight;
   histogram weight / kernel normal;
run;

proc univariate data=sashelp.class;
   class sex;
   var weight;
   histogram weight / kernel normal;
run;
```

The results of these steps are displayed in Figure 1.50, Figure 1.51, Figure 1.52, and Figure 1.53. For more information about kernel density estimation and PROC KDE, see "The KDE Procedure" (*SAS/STAT User's Guide*). For more information about PROC UNIVARIATE, see the *Base SAS Procedures Guide: Statistical Procedures*.

1.5.1   Fringe Plot

A fringe plot consists of small vertical lines that show the location of each individual data value in the bottom portion of the graph. Fringe plots are most frequently overlaid on other graphs such as histograms. The following steps create a histogram, a kernel density function, and a fringe plot:

```
proc template;
   define statgraph classhistden;
      begingraph;
         entrytitle 'Distribution of Weight';
         layout overlay;
            histogram weight;
            densityplot weight / kernel();
            fringeplot weight;
         endlayout;
      endgraph;
   end;
run;

proc sgrender data=sashelp.class template=classhistden;
run;
```

The results are displayed in Figure 1.54. Fringe plots are not available in PROC SGPLOT.

Figure 1.50 PROC KDE Univariate Density

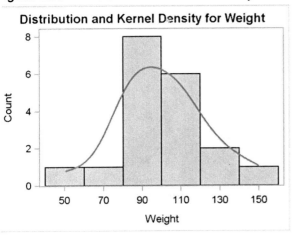

Figure 1.51 PROC KDE Bivariate Density

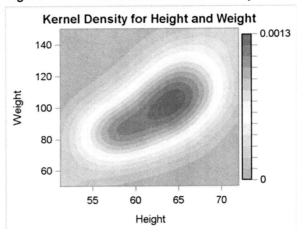

Figure 1.52 PROC UNIVARIATE Histogram

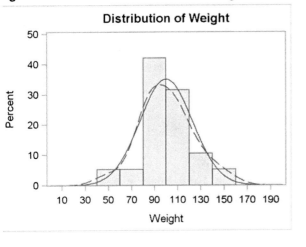

Figure 1.53 PROC UNIVARIATE Grouped Histogram

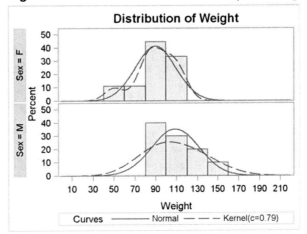

Figure 1.54 Histogram, Density, and Fringe Plot

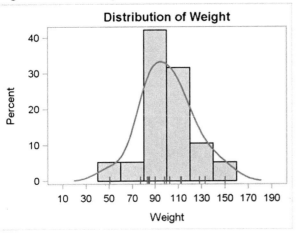

Figure 1.55 Normal Density with Drop Lines

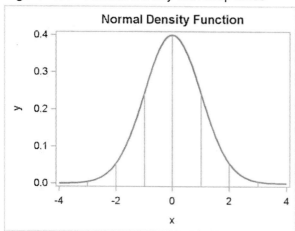

Figure 1.56 Normal Density with Area

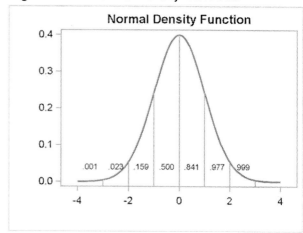

1.5.2 Drop Lines

This section shows how to add a drop line to a plot. A drop line is a line from a point in the graph to one of the axes. The second part of this section also shows some creative uses of expressions. The following steps create a normal density function along with vertical drop lines at –3, –2, –1, 0, 1, 2, and 3:

```
data x(drop=c);
   c = sqrt(2 * constant('pi'));
   do x = -4 to 4 by 0.05;
      y = exp(-0.5 * x ** 2) / c;
      if -3.5 le x le 3.5 and abs(x - round(x)) < 1e-8 then z = y;
      else z = .;
      output;
   end;
run;

proc template;
   define statgraph normal;
      begingraph;
         entrytitle 'Normal Density Function';
         layout overlay;
            seriesplot x=x y=y / lineattrs=graphfit;
            dropline x=x y=z / dropto=x;
         endlayout;
      endgraph;
   end;
run;

proc sgrender data=x template=normal;
run;
```

The DATA step creates the X and Y coordinates of the normal density function, $y = e^{-\frac{1}{2}x^2}/\sqrt{2\pi}$. It also creates a variable z that equals y when x is an integer in the range –3 to 3. The variable z contains the Y axis coordinates of the drop lines and missing values for all other values of the variable x. More precisely, when x is the range –3.5 to 3.5 and when the absolute difference between x and x rounded to the nearest integer is less than 1E–8, a nonmissing value is stored for the variable z.

The template contains a SERIESPLOT statement that draws the normal density function using the **GraphFit** style element. In the STATISTICAL style, this makes a function that is thicker than the default function. The DROPLINE statement draws lines from the X Y coordinates to the X axis due to the DROPTO=X option. The plot is produced by PROC SGRENDER, and the results are displayed in Figure 1.55.

The point of this next example is not to show you how you should do things; rather, it is to show you how you could do things. It relies heavily on expressions in the EVAL function, which has been used in simple ways in preceding examples. This next example uses a data set with only the X and Y coordinates of the normal density function and manufactures all the other information that it needs with expressions. It would have been easier to make all of this information in a DATA step. However, it is instructive to see that you can do it without a DATA step. When you modify a template that SAS provides to use with a SAS analysis procedure, you cannot add new columns to the data object. You can only work with the information that the procedure provides. This example creates a normal density function with drop lines like the preceding example. However it also provides a label for each line that displays the proportion of the area under the curve that falls below the drop line and it suppresses the axis labels. The following steps create the plot that is displayed in Figure 1.56:

```
data x(drop=c);
   c = sqrt(2 * constant('pi'));
   do x = -4 to 4 by 0.05;
      y = exp(-0.5 * x ** 2) / c;
      output;
   end;
run;


proc template;
   define statgraph normal;
      begingraph;
         entrytitle 'Normal Density Function';
         layout overlay;
            seriesplot x=x y=y / lineattrs=graphfit;
            scatterplot x=eval((log(-3.5 le x le 3.5 and
                          abs(x - round(x)) < 1e-8) + 1) * x - 0.5)
                        y=eval(0 * x + 0.04) /
                        markercharacter=eval(put(probnorm(x), 4.3));
            dropline x=x y=eval((log(-3.5 le x le 3.5 and
                          abs(x - round(x)) < 1e-8) + 1) * y) / dropto=x;
         endlayout;
      endgraph;
   end;
run;
```

```
proc sgrender data=x template=normal;
   label x = '00'x y = '00'x;
run;
```

The DATA step creates the variables x and y with the normal density function. The SERIESPLOT statement displays this function and matches the SERIESPLOT statement in the first part of this section. The DROPLINE statement now specifies an expression of the form Y=EVAL(...) rather than a column such as the Y=Z specification shown previously. This expression is as follows:

```
eval((log(-3.5 le x le 3.5 and abs(x - round(x)) < 1e-8) + 1) * y)
```

Part of this expression is as follows:

```
-3.5 le x le 3.5 and abs(x - round(x)) < 1e-8
```

This matches the expression that was used in the IF statement in the previous steps. This Boolean (or logical) expression is true (or 1) when x is an integer in the range –3 to 3 and it is false (or 0) for every other value. For illustration, if you replace this expression by its possible results, "0 or 1", you get the following:

```
eval( (log(0 or 1) + 1) * y )
```

When the Boolean expression is true or 1, then the expression `(log(0 or 1) + 1)` becomes `(log(1) + 1)` which is $0 + 1 = 1$. When the Boolean expression is false or 0, then the expression `(log(0 or 1) + 1)` becomes `(log(0) + 1)` which is missing plus 1 = missing (or "."). The formulation `(log( Boolean expression ) + 1)` evaluates to 1 when the expression is true and missing when it is false. Hence the product, `(log( Boolean expression ) + 1)` * *other expression* becomes the other expression when the Boolean expression is true and missing otherwise. You can use this kind of formulation to select just certain observations for processing in GTL statements that do not provide a FREQ= option or some other mechanism for observation selection. This DROPLINE statement with the expression produces the same results as the simpler DROPLINE statement used in the first part of this section.

The template contains a new statement, a SCATTERPLOT statement, that displays the area under the curve preceding each drop line. The SCATTERPLOT statement is as follows:

```
scatterplot x=eval((log(-3.5 le x le 3.5 and
                     abs(x - round(x)) < 1e-8) + 1) * x - 0.5)
            y=eval(0 * x + 0.04) /
            markercharacter=eval(put(probnorm(x), 4.3));
```

The goal is to place probability values (or areas under the curve) to the left of each line and positioned slightly above the X axis where they will not conflict with the density function. Simplifying like before, the X= option uses the following expression:

```
eval((log(0 or 1) + 1) * x - 0.5)
```

It results in x – 0.5 when the expression is true (x is an integer in the range –3 to 3) and missing otherwise. This expression provides the X coordinates for the scatterplot that displays the area under the curve. The Y axis coordinates are `y=eval(0 * x + 0.04)` and produce constant Y coordinates of 0.04. Simpler expressions like Y=0.04 or Y=EVAL(0.04) are not accepted, so you can specify a constant by adding a constant to 0 times one of the variables. The MARKERCHARACTER= option specifies a label that is displayed in place of a symbol. This variable or expression can be character or numeric, so you could just specify: `MarkerCharacter=Eval(ProbNorm(x))`. However, that would produce more decimal places than you need. The PUT function is used to format the probability or area under the curve to a 4.3 format.

Figure 1.56 has no labels on the axes. This is accomplished by the statement `label x = '00'x y = '00'x` specified in the PROC SGRENDER step. This sets the label for the two variables to null specified as a hexadecimal constant. Alternatively, you could specify the options `xAxisOpts=(Display=(Line Ticks TickValues)) yAxisOpts=(Display=(Line Ticks TickValues))` in the LAYOUT OVERLAY statement.

There are other things that you can do to this graph. Notice that neither Figure 1.55 nor Figure 1.56 display precisely the correct geometry. That is, both have a Y axis where $y = 0$ does not correspond to the X axis. You could specify `yAxisOpts=(OffsetMin=0)` in the LAYOUT OVERLAY statement to correct this. While this fixes the geometry problem it makes the drop lines at −3 and 3 difficult to see.

Figure 1.57   Box Plots with Custom Template

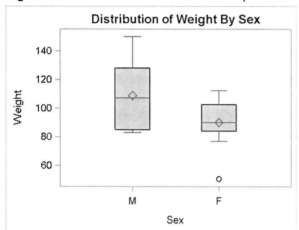

Figure 1.58   Box Plots with PROC SGPLOT

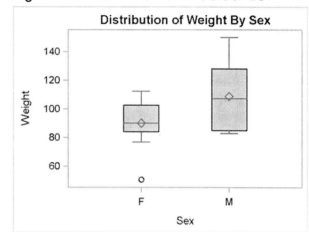

Figure 1.59   Box Plots with Custom Template

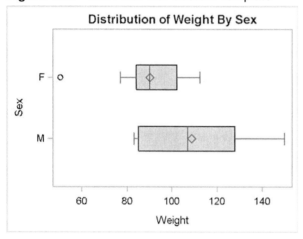

Figure 1.60   Box Plots with PROC SGPLOT

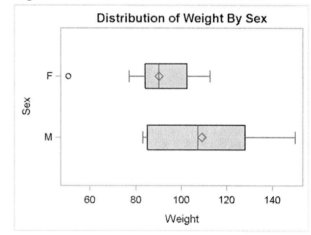

1.6   Box Plot

A box plot displays the distribution of groups of data through their five-number summaries (the minimum, twenty-fifth percentile, median, seventy-fifth percentile, and maximum). Optionally, means and outliers can be displayed as well. The following steps produce vertical box plots of weight by sex:

```
proc template;
   define statgraph classbox;
      begingraph;
         entrytitle 'Distribution of Weight By Sex';
         layout overlay;
            boxplot y=weight x=sex;
         endlayout;
      endgraph;
   end;
run;
```

```
proc sgrender data=sashelp.class template=classbox;
run;

proc sgplot data=sashelp.class noautolegend;
   title 'Distribution of Weight By Sex';
   vbox weight / category=sex;
run;
```

In the GTL, the BOXPLOT statement requests a vertical box plot with the X= option naming the categorical variable and the Y= option naming the quantitative variable. In PROC SGPLOT, the VBOX statement requests a vertical box plot with the quantitative variable specified before the slash and the categorical variable specified in the CATEGORY= option. The results are displayed in Figure 1.57 and Figure 1.58. The results are identical except for the order of the categories. In the GTL, the categories appear in the same order as they do in the input data set. In PROC SGPLOT, the categories appear in sorted order.

The following steps produce horizontal box plots of weight by sex:

```
proc template;
   define statgraph classbox;
      begingraph;
         entrytitle 'Distribution of Weight By Sex';
         layout overlay;
            boxplot y=weight x=sex / orient=horizontal;
         endlayout;
      endgraph;
   end;
run;

proc sgrender data=sashelp.class template=classbox;
run;

proc sgplot data=sashelp.class noautolegend;
   title 'Distribution of Weight By Sex';
   hbox weight / category=sex;
run;
```

The options here are an obvious variation on the options in the previous steps. The results are displayed in Figure 1.59 and Figure 1.60.

You can also use PROC BOXPLOT to produce box plots. The following steps produce vertical and horizontal box plots:

```
ods graphics on;

proc sort data=sashelp.class out=class;
   by sex;
run;

proc boxplot data=class;
   plot weight*sex;
run;
```

Figure 1.61   Vertical Box Plot with PROC BOXPLOT   **Figure 1.62**   Horizontal Box Plot with PROC BOXPL(

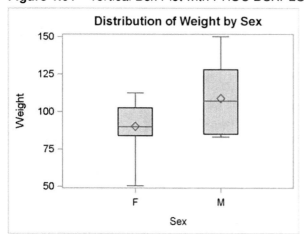   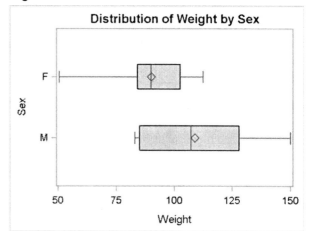

```
proc boxplot data=class;
   plot weight*sex / horizontal;
run;
```

The results are displayed in Figure 1.61 and Figure 1.62. For more information about PROC BOXPLOT, see "The BOXPLOT Procedure" (*SAS/STAT User's Guide*).

The preceding box plot examples provide ODS Graphics or PROC BOXPLOT with raw data and with multiple observations in each group. ODS Graphics or PROC BOXPLOT does the computations to get the statistics and produce the results. The remainder of this section uses only PROC TEMPLATE and PROC SGRENDER with the BOXPLOTPARM statement. This statement requires summarized data, not raw data. You provide a data set with one observation per statistic per group and the information that needs to be displayed. The advantage of using the BOXPLOTPARM statement is you can have precise control over the calculations. You can create this data set from the raw data using a procedure such as PROC UNIVARIATE.

The following steps create the input data set that the BOXPLOTPARM statement requires:

```
proc univariate data=sashelp.class;
   var weight;
   class sex;
   ods output quantiles=q;
run;

data q2;
   set q;
   quantile = scan(quantile, 2, ' ');
   if quantile ne ' ';
run;

proc print;
run;
```

Figure 1.63 Parameterized Vertical Box Plot

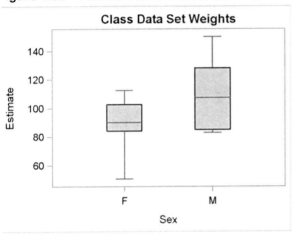

Figure 1.64 Parameterized Horizontal Box Plot

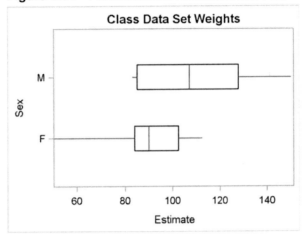

PROC UNIVARIATE is run to get the quantiles (minimum, twenty-fifth percentile, median, seventy-fifth percentile, and maximum) for the variable Weight for each level of the classification variable Sex. The ODS OUTPUT statement is used to create an output data set containing the quantile information. This data set is not in the form that the BOXPLOTPARM statement expects, so a DATA step is used to get the data into the proper form. In the PROC UNIVARIATE table and ODS output data set, there is a variable Estimate with the quantiles and a variable Quantile with the quantile labels, which are as follows: 100% Max, 99%, 95%, 90%, 75% Q3, 50% Median, 25% Q1, 10%, 5%, 1%, 0% Min. Only the Max, Q3, Median, Q1, and Min statistics are needed for a basic box plot, and only those labels are expected (that is without the "$n\%$" part). The DATA step statement `quantile = scan(quantile, 2, ' ')` extracts just the second value where values are delimited by blanks. This creates the labels that the BOXPLOTPARM statement expects. Upper, lower, and mixed case values work equally well. The statement `if quantile ne ' '` discards observations with blank labels leaving the values labeled by: Max, Q3, Median, Q1, and Min. These values are displayed in Figure 1.65. Additional statistics could include: MEAN (data mean), OUTLIER (observation outside the lower and upper fences), FAROUTLIER (observation outside the lower and upper far fences), and STD (standard deviation).

Figure 1.65 Input for a Parameterized Box Plot

```
                      Distribution of Weight By Sex

                      Var
             Obs      Name     Sex     Quantile      Estimate

              1       Weight    F       Max           112.50
              2       Weight    F       Q3            102.50
              3       Weight    F       Median         90.00
              4       Weight    F       Q1             84.00
              5       Weight    F       Min            50.50
              6       Weight    M       Max           150.00
              7       Weight    M       Q3            128.00
              8       Weight    M       Median        107.25
              9       Weight    M       Q1             85.00
             10       Weight    M       Min            83.00
```

The following steps produce the box plot of weight by sex that is displayed in Figure 1.63:

```
proc template;
   define statgraph boxplotparm1;
      begingraph;
         entrytitle "Class Data Set Weights";
         layout overlay;
            boxplotparm y=estimate x=sex stat=quantile;
         endlayout;
      endgraph;
   end;
run;


proc sgrender data=q2 template=boxplotparm1;
run;
```

The BOXPLOTPARM statement specifies the quantile and other statistics in the Y= option, the groups in the X= option, and the labels for the statistics in the STAT= option. The following steps create a horizontal box plot without fill and produce Figure 1.64:

```
proc template;
   define statgraph boxplotparm2;
      begingraph;
         entrytitle "Class Data Set Weights";
         layout overlay;
            boxplotparm y=estimate x=sex stat=quantile /
                        orient=horizontal display=(median);
         endlayout;
      endgraph;
   end;
run;


proc sgrender data=q2 template=boxplotparm2;
run;
```

The option ORIENT=HORIZONTAL creates the horizontal box plot. The default value for the DISPLAY= option is DISPLAY=STANDARD which is the same as DISPLAY=(CAPS FILL MEAN MEDIAN OUTLIERS). The option DISPLAY=(MEDIAN) displays the median in the box, without filling the boxes and without the caps on the lines, or the means or outliers (if they had been provided).

Figure 1.66 Series Plots with Custom Template

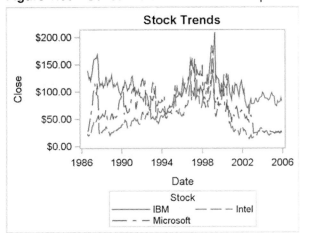

Figure 1.67 Series Plots with PROC SGPLOT

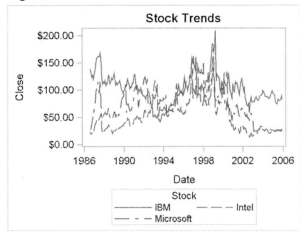

1.7   Series Plot

A series plot is a graphical display of two quantitative variables where the points are connected by straight line segments and are otherwise typically not displayed. When the points are close together, the results can appear to be a smooth curve. Series plots are used to show regression fit functions and trends over time. The following steps produce a series plot for each of three stocks:

```
proc template;
   define statgraph seriesplot;
      begingraph;
         entrytitle 'Stock Trends';
         layout overlay;
            seriesplot x=date y=close / group=stock name='stocks';
            discretelegend 'stocks' / title='Stock';
         endlayout;
      endgraph;
   end;
run;

proc sgrender data=sashelp.stocks template=seriesplot;
run;

proc sgplot data=sashelp.stocks;
   title 'Stock Trends';
   series x=date y=close / group=stock;
run;
```

The SERIESPLOT statement in the GTL and the SERIES statement in PROC SGPLOT connect the data points for X=Date variable on the X axis and Y=Close variable on the Y axis for each value in the categorical GROUP=Stock variable. The SERIESPLOT statement is named so that a legend, entitled 'Stock', is produced. In PROC SGPLOT, the legend is automatically produced. The results are displayed in Figure 1.66 and Figure 1.67.

Figure 1.68 Dot Plot with Custom Template

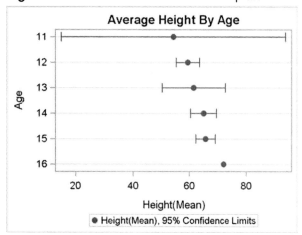

Figure 1.69 Dot Plot with PROC SGPLOT

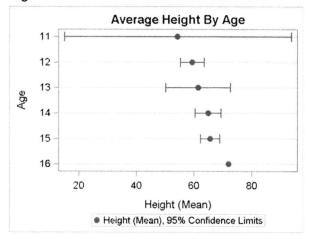

1.8   Dot Plot

A dot plot displays summary statistics computed from a quantitative variable for each level of a classification or group variable. The following steps show two ways to produce dot plots:

```
proc means data=sashelp.class noprint nway;
   var height;
   class age;
   output out=class mean=height lclm=lower uclm=upper;
   label height = 'Height(Mean)';
run;

proc template;
   define statgraph dotplot;
      begingraph;
         entrytitle 'Average Height By Age';
         layout overlay / yaxisopts=(type=discrete griddisplay=on
                                     reverse=true);
            scatterplot y=age x=height / xerrorlower=lower xerrorupper=upper
                        markerattrs=(symbol=circlefilled) name='dot'
                        legendlabel='Height(Mean), 95% Confidence Limits';
            discretelegend 'dot';
         endlayout;
      endgraph;
   end;
run;
```

```
proc sgrender data=class template=dotplot;
run;

proc sgplot data=sashelp.class;
    title 'Average Height By Age';
    dot age / response=height stat=mean limits=both;
run;
```

In PROC SGPLOT, the DOT statement is used to make a dot plot for the RESPONSE=Height variable for the levels of the variable Age, which is specified before the slash. The mean height is shown for each age along with confidence limits in both directions. The results are displayed in Figure 1.69.

The GTL does not have a dot plot statement; however, the SCATTERPLOT statement has all of the needed options once the data have been processed with PROC MEANS. PROC MEANS is used to create an output data set that contains for each age group the mean height and two other variables. The variable Lower contains the lower end of the confidence interval, and the variable Upper contains the upper end of the confidence interval. Additionally, a label is provided for the mean height, and that is the same label that PROC SGPLOT automatically generates. The LAYOUT OVERLAY statement option YAXISOPTS=(TYPE=DISCRETE) specifies that the X axis is discrete (whereas, the default axis is linear). The YAXISOPTS=(REVERSE=TRUE) option reverses the Y axis so that the values decrease as you move vertically up the Y axis instead of increasing.

The SCATTERPLOT statement is used to display the X=Height response variable for the levels of the Y=Age categorical variable. Error bars are drawn in the range from the values in the XERRORLOWER=Lower variable to the values in the XERRORUPPER=Upper variable. The means are displayed as solid filled circles due to the option MARKERATTRS=(SYMBOL=CIRCLEFILLED). The statement is named so that a legend can be produced. A legend label is provided that matches the legend label that PROC SGPLOT automatically displays. The results are displayed in Figure 1.68.

Figure 1.70 Prediction Ellipse with Custom Template

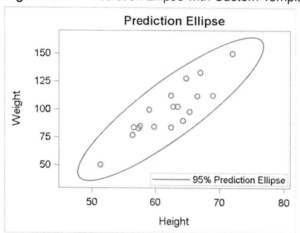

Figure 1.71 Prediction Ellipse with PROC SGPLOT

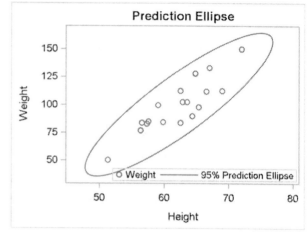

1.9 Ellipses

When the data have a bivariate normal distribution, a confidence ellipse for the population mean or a prediction ellipse for a new observation can be computed and displayed. A prediction ellipse is a region for predicting a new observation in the population. It also approximates a region containing a specified percentage of the population. The ellipse indicates the correlation between the two standardized variables. The following steps produce a scatter plot with a 95% prediction ellipse:

```
proc template;
   define statgraph ellipse;
      begingraph;
         entrytitle 'Prediction Ellipse';
         layout overlay;
            scatterplot x=height y=weight;
            ellipse x=height y=weight / type=predicted
                  legendlabel='95% Prediction Ellipse' name='ellipse';
            discretelegend 'ellipse' / location=inside
                  halign=right valign=bottom;
         endlayout;
      endgraph;
   end;
run;

proc sgrender data=sashelp.class template=ellipse;
run;

proc sgplot data=sashelp.class;
   title 'Prediction Ellipse';
   scatter x=height y=weight;
   ellipse x=height y=weight;
   keylegend / location=inside position=bottomright;
run;
```

In the GTL, the SCATTERPLOT statement is used to display the X=Height response variable for the levels of the Y=Weight categorical variable. The ELLIPSE statement is used to display a 95% prediction ellipse. The ELLIPSE statement is named so that it can appear in the legend, and a legend label is provided. The DISCRETELEGEND statement places the legend inside the plot horizontally placed on the right and vertically placed on the bottom of the plot.

In PROC SGPLOT, the SCATTER statement is used to display the X=Height response variable for the levels of the Y=Weight categorical variable. The ELLIPSE statement is used to display a 95% prediction ellipse. The KEYLEGEND statement is used to place the legend in the inside of the plot in the bottom right. The results are displayed in Figure 1.70 and Figure 1.71.

The two plots differ in that PROC SGPLOT by default places information about both the scatter plot and the ellipse in the legend. However, the graph template only placed the ellipse information in the plot. The following step produces exactly the same plot as is produced using the GTL.

```
proc sgplot data=sashelp.class;
   title 'Prediction Ellipse';
   scatter x=height y=weight;
   ellipse x=height y=weight / name='e';
   keylegend 'e' / location=inside position=bottomright;
run;
```

Since only one statement name is specified in the KEYLEGEND statement, only the legend for the ellipse appears in the plot. The results of this step are not shown.

The preceding ellipse examples provide ODS Graphics with raw data, and ODS Graphics does the computations and produces the results. The remainder of this section uses only PROC TEMPLATE and PROC SGRENDER with the ELLIPSEPARM statement. The ELLIPSEPARM statement requires you to provide all of the parameters for drawing each ellipse.

Figure 1.72   Understanding Ellipses' Slopes

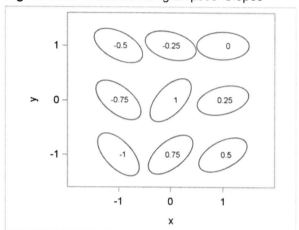

Figure 1.73   Understanding Ellipses

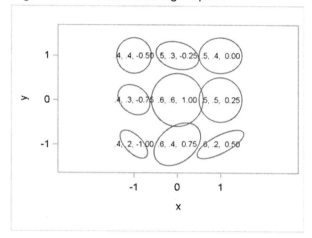

The following steps use the ELLIPSEPARM statement to produce some ellipses just to illustrate how ellipses are defined and how this statement works:

```
data x;
   major = 0.5;
   minor = 0.25;
   input slope x y;
   l = slope;
   datalines;
-1.00  -1 -1
-0.75  -1  0
-0.50  -1  1
-0.25   0  1
 0.00   1  1
 0.25   1  0
 0.50   1 -1
 0.75   0 -1
 1.00   0  0
;

proc template;
   define statgraph ellipse;
      begingraph;
         layout overlayequated / equatetype=square;
            scatterplot x=x y=y / markercharacter=l;
            ellipseparm semimajor=major semiminor=minor slope=slope
                        xorigin=x yorigin=y;
         endlayout;
      endgraph;
   end;
run;

proc sgrender data=x template=ellipse;
run;
```

The results are displayed in Figure 1.72. Ellipses are defined by five parameters, and all must be specified on the ELLIPSEPARM statement. The values can be constant, or they can be variables in a data set. Consider the ellipse in Figure 1.72 that is displayed in the bottom left corner of the plot. It is centered at X=−1 and Y=−1. These X and Y coordinates are the first two parameters. If you were to draw a line through this (X, Y) point to the two points on the ellipse that are farthest from X and Y, the slope of this line would be −1. Half the length of this line is the semimajor parameter value. Half the length of the line that connects two points on the ellipse that is perpendicular to the semimajor line is the semiminor parameter value. When the semimajor parameter equals the semiminor parameter, the ellipse is a circle and the slope is irrelevant. The ellipse centered on the (−1,−1) coordinates has a slope of −1 since that is the slope of the major axis. The length of the major axis is 2 × 0.5, and the length of the minor axis is 2 × 0.25. The shapes of the ellipses in Figure 1.72 differ only in their slopes. Starting in the bottom left corner, and continuing clockwise around and then ending in the middle, the slopes range from −1 to 1.

The LAYOUT statement that is used to make this graph is as follows:

```
layout overlayequated / equatetype=square;
```

This layout is designed to produce a square plot with equated axes. However, plot elements such as ellipses can cause the axes to be padded and thus rectangular, not square. Nevertheless, a centimeter on one axis represents the same data range as a centimeter on the other axis, and both axes have the same ticks and data range. If these axes had not been equated, the geometry of the ellipses would not be properly displayed in the plot. This would be most noticeable in Figure 1.73 (which is created by the next steps) where the circles would be displayed as ellipses. The SCATTERPLOT statement displays the label variable L which contains the slope instead of a marker by using the MARKERCHARACTER= option. The ELLIPSEPARM statement specifies the input data set variables for the five parameters.

The following steps create a series of ellipses with different parameters:

```
data y;
   input slope major minor x y;
   l = put(major, 2.1) || ', ' || put(minor, 2.1) || ', ' || put(slope, 5.2);
   datalines;
-1.00   .4   .2   -1   -1
-0.75   .4   .3   -1    0
-0.50   .4   .4   -1    1
-0.25   .5   .3    0    1
 0.00   .5   .4    1    1
 0.25   .5   .5    1    0
 0.50   .6   .2    1   -1
 0.75   .6   .4    0   -1
 1.00   .6   .6    0    0
;

proc sgrender data=y template=ellipse;
run;
```

The label inside of each ellipse consists of the semimajor parameter, the semiminor parameter, and the slope. The results are displayed in Figure 1.73.

The next section uses the ELLIPSEPARM statement to make a bubble plot.

Figure 1.74 Bubble Plot Based on Area

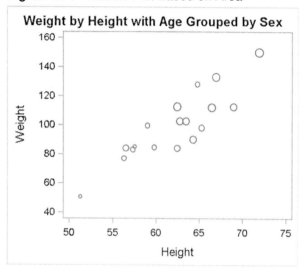

Figure 1.75 Bubble Plot Based on the Radius

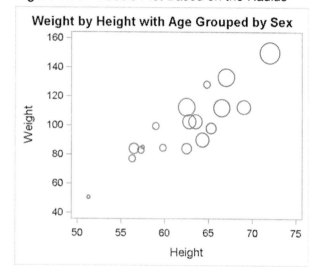

1.10 Bubble Plot

A bubble plot is a scatter plot of two quantitative variables where the symbols are circles and the area or diameter of each circle conveys the value of a third quantitative variable.[3] The GTL and the SG procedures do not yet provide a direct mechanism to make bubble plots. However, with the ELLIPSEPARM statement and a little extra work, you can use PROC TEMPLATE, the GTL, and PROC SGRENDER to make a bubble plot.

Before you can make a bubble plot with the GTL, you must know the range of the X and Y variables, and you must calculate the radius of the circle for each point. This must be done twice, once for each axis variable. Each point has an X coordinate and a Y coordinate, and there is a range of X and Y values. In most cases, the two ranges are different, and X and Y can have different scales. Hence, each bubble is drawn as an ellipse, with a slope of zero, a semimajor parameter corresponding to the radius along the X axis, and a semiminor parameter corresponding to the radius along the Y axis.

The following statements provide the initial specifications for the steps that follow:

```
%let title     = Weight by Height with Age Grouped by Sex;
%let data      = sashelp.class;
%let xvar      = height;
%let yvar      = weight;
%let opts      = group=sex;
%let bubblevar = age - 10;
%let function  = sqrt;
%let scale     = 0.01;
```

---

[3] Some experts suggest that using area is more appropriate than using the radius.

The input in this example is specified through macro variables so that you can specify all of the input at the top. The program is general after the %LET statements, and does not need to be modified for a typical bubble plot. Use the macro variable Title to specify the title, Data to specify the input SAS data set, xVar to specify the X axis variable, yVar to specify the Y axis variable, and Opts to specify options for the ELLIPSEPARM statement. In this case, the GROUP=Sex option displays males differently from females in the plot. With the STATISTICAL style, the females are displayed in red and the males are displayed in blue. Use the macro variable BubbleVar to specify the third variable that is displayed with the bubble. You can also specify an expression. In this case, the ages start at 11, so the expression `age - 10` creates bubbles with areas proportional to 1 for the youngest class members. Use the macro variable Function to specify the function of the bubble variable. Use the square root function when you want the bubble variable's values to be displayed by area, and set the function to blank or null when you want the bubble variable's value to be displayed by the radius. The last parameter is a scale factor that controls the overall bubble size. Use larger scale factors to make larger bubbles.

The following steps compute for each axis the minimum and maximum, the range, the minimum tick, the maximum tick, and the tick increment:

```
proc means data=&data noprint;
   output out=__minmax min(&xvar &yvar)=mx my max(&xvar &yvar)=mxx mxy;
run;

data _null_;
   set __minmax;

   range = max(mxx - mx, 1e-16);
   inc = 10 ** ceil(log10(range) - 1.0);
   if range / inc >= 7.5 then inc = inc * 2;
   if range / inc <= 2.5 then inc = inc / 2;
   if range / inc <= 2.5 then inc = inc / 2;
   call symputx('__xmin', floor(mx / inc) * inc);
   call symputx('__xmax', ceil(mxx / inc) * inc);
   call symputx('__xinc', inc);

   range = max(mxy - my, 1e-16);
   inc = 10 ** ceil(log10(range) - 1.0);
   if range / inc >= 7.5 then inc = inc * 2;
   if range / inc <= 2.5 then inc = inc / 2;
   if range / inc <= 2.5 then inc = inc / 2;
   call symputx('__ymin', floor(my / inc) * inc);
   call symputx('__ymax', ceil(mxy / inc) * inc);
   call symputx('__yinc', inc);
   run;
```

These steps are necessary because you cannot draw circles in a graph unless you know and control the size of the graph and the data range for each axis. You can do this by explicitly setting the minimum and maximum tick for each axis. PROC MEANS computes the minimum (mx and my) and maximum (mxx and mxy) of the X and Y axis variables and stores them in a SAS data set. The DATA step reads these values and computes the ticks. First, the range is computed for the X axis. If the range is less than 1E–16, it is set to 1E–16 to avoid taking a log of zero in the next statement.

The next statement initializes the inc, which is the tick increment. This statement creates an increment that is a power of 10 (for example, 1, 10, 100, .1, .01, and so on). If there are too many ticks, then the increment is multiplied by 2, making increments that are two times a power of 10 (for example, 2, 20, 200, .2, .02, and so on). If there are too few ticks, then the increment is divided by 2 making increments that are 0.5 times a power of 10 (for example, .5, 5, 50, .05, .005, and so on). If there are still too few ticks, then the increment is again divided by 2 making increments that are 0.25 times a power of 10 (for example, .25, 2.5, 25, .025, .0025, and so on).

The minimum tick is the first multiple of the increment that is less than or equal to the minimum. The maximum tick is the first multiple of the increment that is greater than or equal to the maximum. These values are stored in macro variables using the CALL SYMPUTX routine, and the process is repeated for the Y axis. Macro variable names are chosen that begin with two underscores to minimize the chances of those names conflicting with any of your macro variable names.

The following step creates the bubble plot template:

```
proc template;
   define statgraph bubbleplot;
      begingraph;
         entrytitle "&title";
         layout overlay /
            xaxisopts=(linearopts=(
                        viewmin=&__xmin viewmax=&__xmax
                        tickvaluesequence=(start=&__xmin
                        end=&__xmax increment=&__xinc)))
            yaxisopts=(linearopts=(
                        viewmin=&__ymin viewmax=&__ymax
                        tickvaluesequence=(start=&__ymin
                        end=&__ymax increment=&__yinc)));
            scatterplot x=&xvar y=&yvar / datatransparency=1;
            ellipseparm semimajor=__a1 semiminor=__a2 slope=0
                        xorigin=&xvar yorigin=&yvar /
                        outlineattrs=(pattern=solid)
                        &opts;
         endlayout;
      endgraph;
   end;
run;
```

The LAYOUT OVERLAY statement specifies the minimum and maximum tick along with the tick increment for each axis. The same approach is used for the Y axes, so only the X axis options are explained here. The X axis uses the following options to control the axis:

```
xaxisopts=(linearopts=(viewmin=&__xmin viewmax=&__xmax
            tickvaluesequence=(start=&__xmin end=&__xmax increment=&__xinc)))
```

The tick information is specified in the LINEAROPTS=() option within the XAXISOPTS=() option. The starting tick, the stopping tick, and the increment are specified in the START=, END=, and INCREMENT= options. The macro variables that are computed in the DATA step provide the values. These options, however, are not sufficient to control the range of the data along each axis. The options VIEWMIN=&__xMin and VIEWMAX=&__xMax specify the outer range for the X axis. A tick specification of *start* to *end* by *increment* is not guaranteed to produce an axis that begins with *start* or ends with *end* unless those values are also specified on the VIEWMIN= and VIEWMAX= options. Without these options, ticks can be dropped when the data only span a subset of the tick values.

The ELLIPSEPARM statement is used to make the bubble plot. However, it must be overlaid on a nonparameterized plot. Hence a SCATTERPLOT statement with full transparency (DATATRANS-PARENCY=1) is used to overlay the bubble plot on an invisible scatterplot. Alternatively, you can use the SCATTERPLOT statement to place identification information in the center of each bubble. On the ELLIPSEPARM statement, the XORIGIN= and YORIGIN= options are used to specify the coordinates of the center of each circle. The two scatter plot variables (in this case Weight and Height) are specified through macro variables. The slope is not relevant with circles, so a slope of zero is specified. The options SEMIMAJOR= and SEMIMINOR= specify the two parameters of the ellipse that are required to form the circle. They are computed in the next step and are added as variables to the input SAS data set.

The five required parameters of the ellipse are specified in the ELLIPSEPARM statement before the slash. The parameters of an ellipse are explained in detail starting on page 54. After the slash, any additional options are specified with the &Opts macro variable. In addition, the option `OutLineAttrs=(pattern=solid)` is used to draw solid circles. When there are multiple groups, the line patterns from the `GraphData`*n* style elements are used which specify various patterns of dashed lines for all but the first circle.

The following steps compute the semimajor and semiminor parameters and create the plot:

```
data __minmax;
   set &data;
   __a1 = &scale * (&__xmax - &__xmin) * &function(max(&bubblevar, 1e-16)) *
          (480 / 640);
   __a2 = &scale * (&__ymax - &__ymin) * &function(max(&bubblevar, 1e-16));
run;

proc sgrender data=__minmax template=bubbleplot;
run;
```

The results are displayed in Figure 1.73. This plot displays a quantitative variable Weight on the Y axis, a quantitative variable Height on the X axis, a quantitative variable Age in the bubble size, and a qualitative variable Sex though the bubble color differences.

The semimajor and semiminor parameters are calculated for each point in the input data set and stored in the variables __a1 and __a2. Variable names are chosen that begin with two underscores to minimize the chances of those names conflicting with any of your variable names. Consider the variable __a2 first. It is computed by first taking the maximum of the bubble variable expression and 1E–16 to ensure that invalid data are not passed to the square root function. The value of the macro variable Function is SQRT. Since area is π times the radius squared, and since we want bubble area to be related to the bubble variable, we take the square root of the bubble variable expression. This is multiplied by the range of values on the Y axis. This product is then multiplied by a scale factor (in this case 0.01) to fine tune the bubble sizes. You can modify this scale factor to uniformly stretch or shrink the bubbles. The calculations for the X axis are the same as those for the Y axis except for one additional multiplication. By default, graphs are designed for a size of 480 pixels high and 640 pixels wide. The X axis computations are adjusted by the proportion $480/640 = 3/4$ to account for the differing lengths of the axes. This adjustment is based on the outer box size of 480 by 640, not the unknown graph size, so it is approximate and may need to be adjusted particularly for smaller graphs. PROC SGRENDER uses the custom template and the new data set with the original data and the ellipse parameters to make the graph.

The following parameters differ from the previous ones only in that the square root function was removed from the value of the macro variable Function:

```
%let title    = Weight by Height with Age Grouped by Sex;
%let data     = sashelp.class;
%let xvar     = height;
%let yvar     = weight;
%let opts     = group=sex;
%let bubblevar = age - 10;
%let function = ;
%let scale    = 0.01;
```

Using this set of specifications and the same code as before, the results are displayed in Figure 1.75. Since the square root transformation is not used, the circles are bigger and grow at a faster rate in Figure 1.75 than they do in Figure 1.74.

The template in this example could have instead been written as follows:

```
proc template;
   define statgraph bubbleplot;
      nmvar __xmin __xmax __xinc __ymin __ymax __yinc;
      mvar title xvar yvar;
      begingraph;
         entrytitle title;
         layout overlay /
            xaxisopts=(linearopts=(
                      viewmin=__xmin viewmax=__xmax
                      tickvaluesequence=(start=__xmin
                      end=__xmax increment=__xinc)))
            yaxisopts=(linearopts=(
                      viewmin=__ymin viewmax=__ymax
                      tickvaluesequence=(start=__ymin
                      end=__ymax increment=__yinc)));
            scatterplot x=xvar y=yvar / datatransparency=1;
            ellipseparm semimajor=__a1 semiminor=__a2 slope=0
                      xorigin=xvar yorigin=yvar /
                      outlineattrs=(pattern=solid)
                      &opts;
         endlayout;
      endgraph;
   end;
run;
```

This version uses the NMVAR statement to name the macro variables whose values are numeric and the MVAR statement to name the macro variables whose values are character. The ampersands are removed from the macro variable names, and their values are retrieved at PROC SGRENDER run time rather than before the template is compiled by PROC TEMPLATE. Note, however, that this approach cannot be completely implemented here due to the Opts macro variable which can contain one or more option names along with their values. MVAR, NMVAR, and DYNAMIC variables cannot be used in this way; hence the ampersand must be used with Opts, and the template must be recompiled for each new bubble plot.

Figure 1.76 Vertical Line Plot with Custom Template

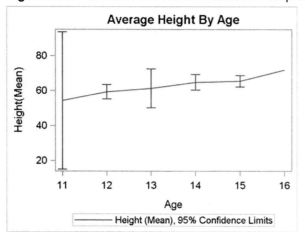

Figure 1.77 Vertical Line Plot with PROC SGPLOT

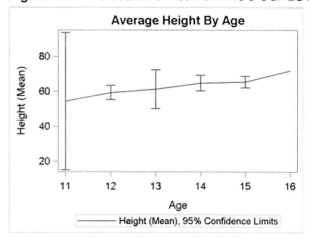

Figure 1.78 Horizontal Line Plot with Custom Template

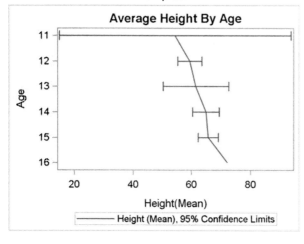

Figure 1.79 Horizontal Line Plot with PROC SGPLOT

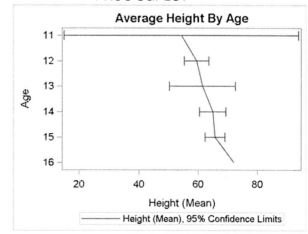

1.11 Line Plot

A line plot displays summary statistics computed from a quantitative variable for each level of an ordered categorical variable. The points in the plot are connected by line segments. The following steps produce a vertical line plot with 95% confidence limits:

```
proc means data=sashelp.class noprint nway;
   var height;
   class age;
   output out=class mean=height lclm=lower uclm=upper;
   label height = 'Height(Mean)';
run;
```

```
proc template;
   define statgraph vline;
      begingraph;
         entrytitle 'Average Height By Age';
         layout overlay / xaxisopts=(type=discrete);
            seriesplot x=age y=height /
               legendlabel='Height (Mean), 95% Confidence Limits' name='vline';
             scatterplot x=age y=height / markerattrs=(size=0)
               yerrorupper=upper yerrorlower=lower;
            discretelegend 'vline';
         endlayout;
      endgraph;
   end;
run;

proc sgrender data=class template=vline;
run;

proc sgplot data=sashelp.class;
   title 'Average Height By Age';
   vline age / response=height stat=mean limits=both;
run;
```

In PROC SGPLOT, the VLINE statement is used to make a vertical line plot for the RE-SPONSE=Height variable for the levels of the variable Age that is specified before the slash. The mean height is shown for each age along with confidence limits in both directions. The results are displayed in Figure 1.77.

The GTL does not have a vertical line plot statement. However, the SCATTERPLOT and SERIES-PLOT statements have all of the needed options once the data have been processed with PROC MEANS. PROC MEANS is used to create an output data set that contains for each age group the mean height and two other variables. The variable Lower contains the lower end of the confidence interval, and the variable Upper contains the upper end of the confidence interval. Additionally, a label is provided for the mean height, and that is the same label that PROC SGPLOT automatically generates. The LAYOUT OVERLAY statement option XAXISOPTS=(TYPE=DISCRETE) specifies that the X axis is discrete (whereas, the default axis is linear). The SCATTERPLOT statement is used to display the X=Height response variable for the levels of the Y=Age categorical variable. However, the points themselves are not displayed due to the specification MARKERATTRS=(SIZE=0). Error bars are drawn in the range from the values in the YERRORLOWER=Lower variable to the values in the YERRORUPPER=Upper variable. The means are displayed as a series plot. No bar is displayed for the mean age of 16 since there is only one observation for that age group. The SERIESPLOT statement is named so that a legend can be produced. A legend label is provided that matches the legend label that PROC SGPLOT automatically displays. The results are displayed in Figure 1.76.

The following steps produce a horizontal line plot with 95% confidence limits:

```
proc means data=sashelp.class noprint nway;
   var height;
   class age;
   output out=class mean=height lclm=lower uclm=upper;
   label height = 'Height(Mean)';
run;

proc template;
   define statgraph hline;
      begingraph;
         entrytitle 'Average Height By Age';
         layout overlay / yaxisopts=(type=discrete reverse=true);
            seriesplot x=height y=age /
               legendlabel='Height (Mean), 95% Confidence Limits'
                          name='hline';
              scatterplot x=height y=age / markerattrs=(size=0)
                 xerrorupper=upper xerrorlower=lower;
            discretelegend 'hline';
         endlayout;
      endgraph;
   end;
run;

proc sgrender data=class template=hline;
run;

proc sgplot data=sashelp.class;
   title 'Average Height By Age';
   hline age / response=height stat=mean limits=both;
run;
```

The steps that produce the horizontal line plots are obvious variations on the steps that produce the vertical line plots. The results are displayed in Figure 1.78 and Figure 1.79.

Figure 1.80 Needle Plot with Custom Template

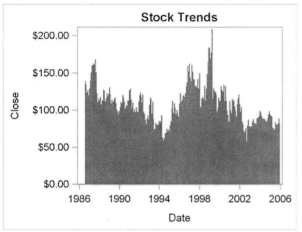

Figure 1.81 Needle Plot with PROC SGPLOT

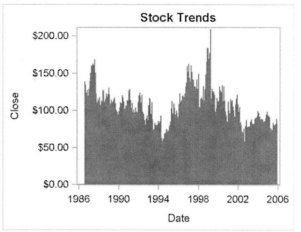

1.12 Needle Plot

A needle plot is a scatter plot where each point is displayed as a line connecting the horizontal axis and the coordinates of the point. The following steps produce a needle plot:

```
proc template;
   define statgraph needle;
      begingraph;
         entrytitle 'Stock Trends';
         layout overlay;
            needleplot x=date y=close;
         endlayout;
      endgraph;
   end;
run;

proc sgrender data=sashelp.stocks template=needle;
run;

proc sgplot data=sashelp.stocks;
   title 'Stock Trends';
   needle x=date y=close;
run;
```

In the GTL, the NEEDLEPLOT statement produces a needle plot for the X=Date variable on the X axis and Y=Close variable on the Y axis. In PROC SGPLOT, the NEEDLE statement produces a needle plot using the same options. The results are displayed in Figure 1.80 and Figure 1.81. With sparser data sets, the distinct lines are visible.

Figure 1.82 Step Plot with Custom Template

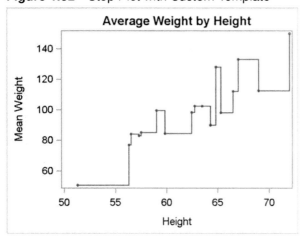

Figure 1.83 Step Plot with PROC SGPLOT

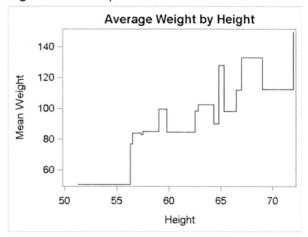

1.13   Step Plot

A series plot is a graphical display of two quantitative variables where the points are connected by a combination of horizontal and vertical lines. The points are optionally displayed on the lines. The function is like a set of stairs which has horizontal landings and vertical risers but no diagonal lines. You can choose to have the vertical segment rise from a point, to a point, or in the middle between points. The following steps produce a step plot:

```
proc means data=sashelp.class noprint nway;
   var weight;
   class height;
   output out=class mean=weight lclm=lower uclm=upper;
   label weight = 'Mean Weight';
run;

proc template;
   define statgraph step;
      begingraph;
         entrytitle 'Average Weight by Height';
         layout overlay;
            stepplot x=height y=weight / display=(markers)
                                         markerattrs=(size=3px);
         endlayout;
      endgraph;
   end;
run;

proc sgrender data=class template=step;
run;
```

```
proc sgplot data=class;
   title 'Average Weight by Height';
   step x=height y=weight;
run;
```

In the GTL, the STEPPLOT statement connects the data points for the X=Height variable on the X axis and Y=Weight variable on the Y axis. The results are displayed in Figure 1.82. In PROC SGPLOT, the STEP statement connects the data points using the same options. The results are displayed in Figure 1.83.

The plots differ because the options DISPLAY=(MARKERS) MARKERATTRS=(SIZE=3PX) are specified in the GTL. These options display the markers at the means using a 3 pixel symbol. The default marker in the STATISTICAL style is a 5 pixel circle. The following step uses PROC SGPLOT to create the same plot that was created by PROC SGRENDER.

```
proc sgplot data=class;
   title 'Average Height by Weight';
   step x=height y=weight / markers markerattrs=(size=3px);
run;
```

The results of this step are not shown, but they match Figure 1.82.

Figure 1.84 Vector Plot with Custom Template

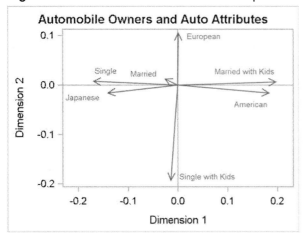

Figure 1.85 Vector Plot with PROC SGPLOT

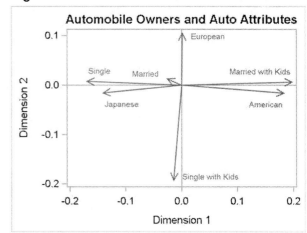

1.14 Vector Plot

A vector plot is a graphical display of two quantitative variables where each point is represented by a vector or line emanating from the origin and terminating at the point. A reference line is a horizontal or vertical line running the width of length of the graph. The following steps produce a vector plot with reference lines from some correspondence analysis results:

```
data corresp;
   input Type $ Name $ 5-25 Dim1 Dim2;
   label dim1 = 'Dimension 1' dim2 = 'Dimension 2';
   datalines;
OBS  Married              -0.02783     0.01339
OBS  Married with Kids     0.19912     0.00639
OBS  Single               -0.17160     0.00762
OBS  Single with Kids     -0.01440    -0.19470
VAR  American              0.18472    -0.01660
VAR  European              0.00129     0.10734
VAR  Japanese             -0.14278    -0.01630
;

proc template;
   define statgraph vector;
      begingraph;
         entrytitle 'Automobile Owners and Auto Attributes';
         layout overlayequated / equatetype=fit;
            referenceline x=0;
            referenceline y=0;
            vectorplot y=dim2 x=dim1 xorigin=0 yorigin=0 /
                        datalabel=name group=type lineattrs=(pattern=solid);
         endlayout;
      endgraph;
   end;
run;
```

```
proc sgrender data=corresp template=vector;
run;

proc sgplot data=corresp noautolegend;
   title 'Automobile Owners and Auto Attributes';
   refline 0 / axis=x;
   refline 0 / axis=y;
   vector x=dim1 y=dim2 / datalabel=name group=type lineattrs=(pattern=solid);
run;
```

The GTL REFERENCELINE statement options and the PROC SGPLOT REFLINE statements produce reference lines along the X and Y axes at 0. The reference line statements are specified ahead of the vector plot statements so that the reference lines are drawn first, and the vectors next. This way if there are any collisions, the reference lines are in the background and are covered by the information in the vector plot.

In the GTL, the VECTORPLOT statement produces vectors, emanating from the origin (XORIGIN=0 YORIGIN=0) and ending at the coordinates specified by the X=Dim1 variable on the X axis and Y=Dim2 variable on the Y axis. The vectors are labeled by the values of the DATALABEL=Name variable. There are two groups of observations that are distinguished by the GROUP=Type variable. The LINEATTRS=(PATTERN=SOLID) option produces solid vectors. By default with the STATISTICAL style, the vectors use the group line styles, and the red vector is dashed. The LAYOUT statement is as follows:

```
layout overlayequated / equatetype=fit;
```

This layout produces a plot with equated axes, where a centimeter on one axis represents the same data range as a centimeter on the other axis. The lengths of the axes are stretched to fit in the outer box. The results are displayed in Figure 1.84.

In PROC SGPLOT, the VECTOR statement produces vectors, emanating from the origin and ending at the coordinates specified by the X=Dim1 variable on the X axis and Y=Dim2 variable on the Y axis. The vectors are labeled by the values of the DATALABEL=Name variable. There are two groups of observations that are distinguished by the GROUP=Type variable. The LINEATTRS=(PATTERN=SOLID) option produces solid vectors. By default with the STATISTICAL style, the vectors use the group line styles, and the red vector is dashed. The results are displayed in Figure 1.85.

The two plots are similar but slightly different. The difference is due to the equated axes in Figure 1.84 versus the standard nonequated axes in Figure 1.85. PROC SGPLOT does not support equated axes.

Figure 1.86   Contour of Swirl Data

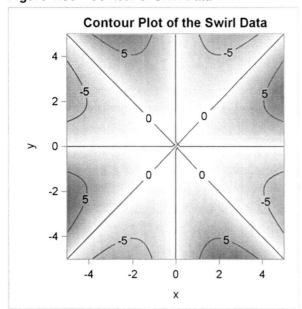

Figure 1.87   Contour of Bivariate Normal Density

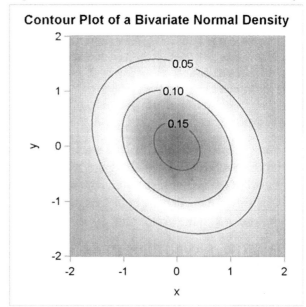

1.15   Contour Plot

A contour plot displays three quantitative variables. The X and Y axis variables form a regular and rectangular grid and the values of the Z variable is displayed with shades of colors that vary over the range of Z. A contour plot can be used to show probability density functions, surfaces, and terrain. The following steps create contour plots of the surface $z = xy(x^2 - y^2)/(x^2 + y^2)$ (with $z = 0$ when $x = 0$ and $y = 0$) and a bivariate normal density function:

```
proc template;
   define statgraph contour;
      mvar title;
      begingraph;
         entrytitle title;
         layout overlayequated / equatetype=square
                xaxisopts=(offsetmin=0 offsetmax=0)
                yaxisopts=(offsetmin=0 offsetmax=0);
            contourplotparm x=x y=y z=z;
         endlayout;
      endgraph;
   end;
run;
```

```
data swirl;
   do x = -5 to 5 by 0.1;
      do y = -5 to 5 by 0.1;
         z = x * x + y * y;
         if z > 1e-16 then z = x * y * (x * x - y * y) / z;
         output;
      end;
   end;
run;

data normal;
   do x = -2 to 2 by 0.1;
      do y = -2 to 2 by 0.1;
         z = 0.164 * exp(-0.5 * ((x + y * 0.25) * x + (x * 0.25 + y) * y));
         output;
      end;
   end;
run;

%let title = Contour Plot of the Swirl Data;
proc sgrender data=swirl template=contour;
run;

%let title = Contour Plot of a Bivariate Normal Density;
proc sgrender data=normal template=contour;
run;
```

The results are displayed in Figure 1.86 and Figure 1.87. Both plots are created by the GTL and PROC SGRENDER. While PROC SGPLOT and the other SG procedures create a rich variety of plots, they cannot be used to create every plot that you can create with the GTL. In particular, the SG procedures cannot make contour plots. Also, PROC SGPLOT cannot be used to make equated axes where a centimeter on one axis represents the same data range as a centimeter on the other axis.

This template has a LAYOUT OVERLAYEQUATED statement with the option EQUATE-TYPE=SQUARE. This creates an equated plot that is square. By default with the LAYOUT OVERLAYEQUATED statement, the axes are equated but the plot is rectangular. The options `xAxisOpts=(OffsetMin=0 OffsetMax=0) yAxisOpts=(OffsetMin=0 OffsetMax=0)` are used to ensure that no extra white space appears between the edges of the contours and the axes. By default, a small offset or white space might appear on the axes. The CONTOURPLOTPARM statement specifies the X axis variable X and the Y axis variable Y along with the density variable Z. An MVAR statement is used to set the title at the time that PROC SGRENDER is run instead of the time that PROC TEMPLATE is run.

Figure 1.88 Contour of Swirl Data   **Figure 1.89** Contour of Bivariate Normal Density

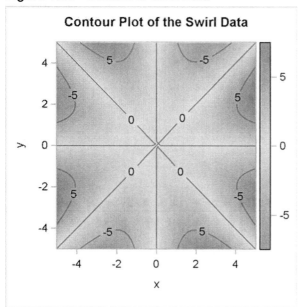

1.15.1   Continuous Legend

This example is a continuation of the previous example. The following steps create contour plots with a continuous legend using a customized style:

```
proc template;
   define style mystyle;
      parent = Styles.statistical;
      class ThreeColorRamp /
         endcolor = CX6666FF
         neutralcolor = CXFFBBFF
         startcolor = CXFF6666
      end;
   end;

   define statgraph contour;
      mvar title;
      begingraph;
         entrytitle title;
         layout overlayequated / equatetype=square
               xaxisopts=(offsetmin=0 offsetmax=0)
               yaxisopts=(offsetmin=0 offsetmax=0);
            contourplotparm x=x y=y z=z / name='cont';
            continuouslegend 'cont';
         endlayout;
      endgraph;
   end;
run;

ods listing style=mystyle;
```

```
%let title = Contour Plot of the Swirl Data;
proc sgrender data=swirl template=contour;
run;

%let title = Contour Plot of a Bivariate Normal Density;
proc sgrender data=normal template=contour;
run;

ods listing;
```

The results are displayed in Figure 1.88 and Figure 1.89. The continuous legend is created by the CONTINUOUSLEGEND statement with the name 'cont', which is also specified in the CONTOURPLOTPARM statement in the NAME= option. The continuous legend is displayed by default vertically on the right side of the plot. It shows the mapping between the values of a variable and a range of colors, in this case from red to magenta to blue. You can also use continuous legends with scatter plots and other types of graphs. For example, you can create a scatter plot of two variables, y and x, on the Y and X axes and use color to show a third variable, z.

The colors in Figure 1.88 and Figure 1.89 differ from the colors in Figure 1.90 and Figure 1.91 due to the redefinition of the `ThreeColorRamp` style element, which by default controls the colors of contour plots. A new ODS style, MYSTYLE, is created using the statement: `define style mystyle`. This style inherits its properties from the STATISTICAL style due to the option PARENT=STYLES.STATISTICAL. The CLASS statement provides the new color ramp definition and the new colors. Colors are specified in values of the form CX*rrggbb*, where the last six characters specify RGB (red, green, blue) values on the hexadecimal scale of 00 to FF (or 0 to 255 base 10). The start, neutral, and end colors are set to CXFF6666 (red, but closer to white than the pure red color CXFF0000), CXFFBBFF (magenta, but closer to white than the pure magenta color CXFF00FF), CX6666FF (blue, but closer to white than the pure blue color CX0000FF), respectively. Note that the shade of magenta used here has more white in it than either the shade of red or the blue used here. The new style is specified in the ODS LISTING statement with the STYLE= option. At the end, the ODS LISTING statement without the STYLE= option restores the default style.

This is a typical way to make a style change. You create a new style that inherits from an existing style and just changes one or a few things. The default for the COLORMODEL= option in the CONTOURPLOTPARM statement is COLORMODEL=`ThreeColorRamp`, which is how you know what style element to change. You learn how to change it by submitting the following step:

```
proc template;
   source styles.default;
   source styles.statistical;
run;
```

Search the results from the bottom up for the definition of `ThreeColorRamp` and copy and replace it with the definition that you want. If you do not know the inheritance structure of your style, just use one SOURCE statement, find the parent style (if any), and repeat the process until you find the final parent. Most styles that are used with statistical graphics inherit from the DEFAULT style. Some elements are defined indirectly. For example, the definition of the end color is: `endcolor = GraphColors('gramp3cend')`. A subsequent search shows that `'gramp3cend'` = `cx667FA2`, which is a shade of blue.

Figure 1.90   Contour of Bivariate Normal Density

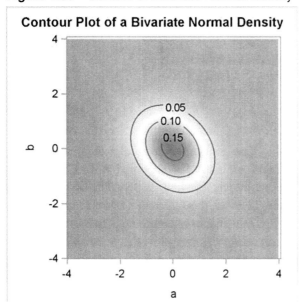

Figure 1.91   Bivariate Normal Density and Sample

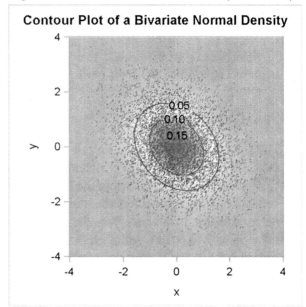

1.15.2   Contour Plot and Scatter Plot Overlaid

This example is a continuation of the previous example and makes several important points about template writing. In this first part, a SCATTERPLOT statement is added in front of the CONTOUR-PLOTPARM statement. You will see that this particular specification overlays the scatter plot and the contour plot but not in the most desirable way. The following steps create Figure 1.90:

```
proc template;
   define statgraph contourplotparm;
      mvar title;
      begingraph;
         entrytitle title;
         layout overlayequated / equatetype=square
            xaxisopts=(offsetmin=0 offsetmax=0)
            yaxisopts=(offsetmin=0 offsetmax=0);
            scatterplot x=a y=b / markerattrs=(size=1px);
            contourplotparm x=x y=y z=z;
         endlayout;
      endgraph;
   end;
run;

data normal;
   do x = -4 to 4 by 0.1;
      do y = -4 to 4 by 0.1;
         z = 0.164 * exp(-0.5 * ((x + y * 0.25) * x + (x * 0.25 + y) * y));
         output;
      end;
   end;
run;
```

```
data sample(drop=i);
   do i = 1 to 10000;
      a = normal(104);
      b = -0.25 * a + sqrt(1 - 0.25 ** 2) * normal(104);
      output;
   end;
run;

data normal;
   merge normal sample;
run;


proc sgrender data=normal template=contourplotparm;
run;
```

In this example, there are four variables that are plotted, X and Y (as in the previous example) and also A and B, which contain a random sample from the bivariate normal distribution. The plot displayed in Figure 1.90 has 'a' and 'b' (the scatter plot variable names) on the axes and has white space between the contours and the axes. However, the scatter plot itself is not displayed. While the GTL syntax used in this example is perfectly valid, it does not make the most desirable graph for several reasons that are discussed in the rest of this section. The graph that displays both the scatter plot and the contour plot is displayed in Figure 1.91 and is produced by the following steps:

```
proc template;
   define statgraph contourplotparm;
      mvar title;
      begingraph;
         entrytitle title;
         layout overlayequated / equatetype=square
            commonaxisopts=(viewmin=-4 viewmax=4)
            xaxisopts=(offsetmin=0 offsetmax=0)
            yaxisopts=(offsetmin=0 offsetmax=0);
            contourplotparm x=x y=y z=z / contourtype=gradient;
            scatterplot x=a y=b / markerattrs=(size=1px);
            contourplotparm x=x y=y z=z / contourtype=labeledline;
         endlayout;
      endgraph;
   end;
run;

data normal;
   do x = -4 to 4 by 0.1;
      do y = -4 to 4 by 0.1;
         z = 0.164 * exp(-0.5 * ((x + y * 0.25) * x + (x * 0.25 + y) * y));
         output;
      end;
   end;
run;

data sample(drop=i);
   do i = 1 to 10000;
      a = normal(104);
      b = -0.25 * a + sqrt(1 - 0.25 ** 2) * normal(104);
      output;
   end;
run;
```

```
data normal;
   merge normal sample;
run;

proc sgrender data=normal template=contourplotparm;
run;
```

The graph in Figure 1.90 is produced by first creating a scatter plot (since the SCATTERPLOT statement is first) and then by creating a contour plot on top of the scatter plot. Statements are executed in the order that they are specified in the template. The scatter plot is created in Figure 1.90, but it is completely overwritten by the contour plot. In Figure 1.91, the contour plot is written first so the scatter plot is visible, and then the contour lines are written on top over both the contour plot and the scatter plot. In Figure 1.90, the SCATTERPLOT statement is the primary plot statement since it is specified first in the layout. The default axis features (axis type, axis labels, and so on) are set by the primary plot. When the statement order is reversed to create Figure 1.91, the CONTOURPLOTPARM statement becomes primary, and the axis labels are 'x' and 'y' instead of 'a' and 'b'. Alternatively, any plot statement can be designated as primary by specifying the option PRIMARY=TRUE after a slash.

Figure 1.90 is produced using the options `xAxisOpts=(OffsetMin=0 OffsetMax=0)` `yAxisOpts=(OffsetMin=0 OffsetMax=0)` to eliminate white space between the data values and the axes. In Figure 1.91, the options `CommonAxisOpts=(ViewMin=-4 ViewMax=4)` are additionally specified. In Figure 1.90, there is some additional white space between the axes and the contour plot near the maximum value of 4 on both axes. This does not happen in Figure 1.91. The difference occurs because the random sample contains values slightly outside the range of –4 to 4, which is used for the population distribution. The options `CommonAxisOpts=(ViewMin=-4` `ViewMax=4)` that are specified for Figure 1.91 ensure that these values are not displayed and do not add additional white space. These options are also useful when there are outliers that compress the display of other points.

The option `MarkerAttrs=(Size=1Px)` is used to create 1-pixel markers. The scatter plot consists of 10,000 points, and larger markers (such as the default 7-pixel markers) are too large to appropriately display both the sample and the population.

The data set Normal has 5 variables. The variables X, Y, and Z contain 6561 nonmissing values, whereas the variables A, and B contain 10,000 nonmissing values. Hence the variables X, Y, and Z contain a block of 3439 missing values. This is not a problem, and it is not unusual for some variables to have a large block of missing values when different plots are overlaid.

There are two basic structures for the input data sets that are used when the graphical display consists of multiple components. Graphs such as those shown in Figure 1.20, Figure 1.21, Figure 1.66, Figure 1.67 and elsewhere use a GROUP= variable to display multiple groups of observations in the same plot. Data from different groups are stacked in the same variables, and a single plotting statement is used in the template. In contrast, and in this example, different graph types are produced, multiple plotting statements are used, and separate variables in the input data set are used to make each plot.

Figure 1.92 Block Plot with Annual Cycle

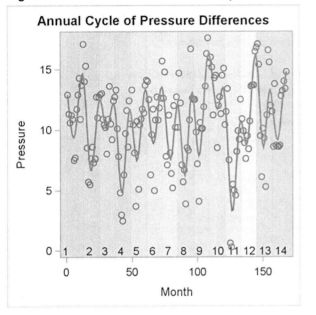

Figure 1.93 Block Plot with El Niño Cycle

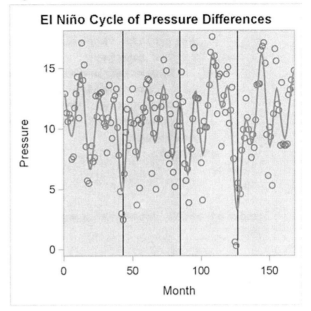

1.16 Block Plot

A block plot visually enhances an overlaid graph by providing a set of colors that highlight and distinguish intervals of one of the plotted variables. The following steps create a fit plot overlaid on a block plot:

```
data ENSO;
   input Pressure @@;
   Month  = _N_;
   Year   = ceil(month / 12);
   ElNino = ceil(month / 42);
   format Pressure 4.1;
   format Month 3.0;
   datalines;
12.9  11.3  10.6  11.2  10.9   7.5   7.7  11.7
12.9  14.3  10.9  13.7  17.1  14.0  15.3   8.5
 5.7   5.5   7.6   8.6   7.3   7.6  12.7  11.0
12.7  12.9  13.0  10.9  10.4  10.2   8.0  10.9

   ... more lines ...

;
```

```
proc template;
   define statgraph block;
      begingraph;
         entrytitle 'Annual Cycle of Pressure Differences';
         layout overlay;
            blockplot x=month  block=year / valuevalign=bottom
               datatransparency=0.75 display=(fill values);
            scatterplot   y=pressure x=month;
            pbsplineplot y=pressure x=month;
         endlayout;
      endgraph;
   end;
run;


proc sgrender data=enso template=block;
run;
```

The results are displayed in Figure 1.92. The ENSO data set contains a variable, Year, that contains integer values in the range 1 to 14. The block plot displays each year using a color that is distinguishable from the other colors in the vicinity (although colors can eventually repeat). The X axis variable is month, which is the same as in the scatter plot and penalized B-spline plot. The values of the Year variable are displayed since the DISPLAY= option specifies VALUES. They are displayed at the bottom of the graph since the option VALUEVALIGN=BOTTOM is specified. By default, the standard display is unoutlined filled bands without labels. The locations where the values are displayed are controlled by the options VALUEHALIGN=LEFT | CENTER | RIGHT | START and VALUEVALIGN=TOP | CENTER | BOTTOM.

The following steps create a fit plot overlaid on a block plot of the 42-month El Niño cycle, which is displayed in Figure 1.93:

```
proc template;
   define statgraph block2;
      begingraph;
         entrytitle 'El Ni(*ESC*){Unicode "00F1"x}o '
                    'Cycle of Pressure Differences';
         layout overlay / xaxisopts=(offsetmin=0 offsetmax=0);
            blockplot x=month  block=elnino /
               datatransparency=0.75 display=(fill outline);
            scatterplot   y=pressure x=month;
            pbsplineplot y=pressure x=month;
         endlayout;
      endgraph;
   end;
run;


proc sgrender data=enso template=block2;
run;
```

In Figure 1.92, the values are displayed along with a filled area for each year due to the option DISPLAY=(FILL VALUES). The boxes on both ends of the X axis are a little larger than the other boxes. In Figure 1.93, the values are not displayed, and an outlined filled area for each cycle is displayed due to the option DISPLAY=(FILL OUTLINE). All boxes are the same width due to the specification: **xAxisOpts=(OffsetMin=0 OffsetMax=0)**. There is a trade-off here. In Figure 1.92, the boxes are not all uniform, but there is room on both edges for the markers, which are centered at the beginning and end of each year and hence extend beyond the year. In Figure 1.93, the boxes are uniform, but the extreme markers are clipped. In both graphs, the option DATATRANSPARENCY=0.75 is specified so that the boxes are 75% as transparent as the default. This makes lighter boxes and a more subtle effect. In Figure 1.93, the unicode specification **(*ESC*){Unicode "00F1"x}** creates the letter "ñ" in "El Niño" in the title. A quoted string followed by an "x" means the value is interpreted as a hexadecimal constant. The Unicode Consortium (http://unicode.org/) provides a page of character codes at http://www.unicode.org/charts/charindex.html.

Figure 1.94 Surface Plot

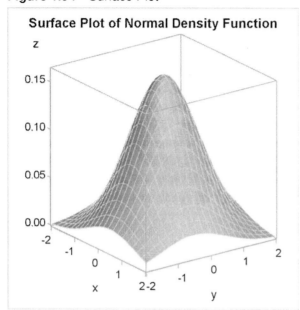

Figure 1.95 Wire-Frame Surface Plot

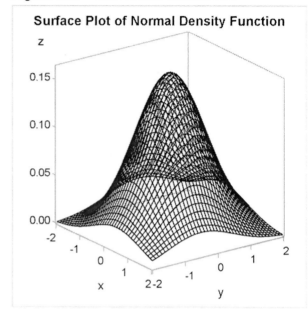

1.17   Three-Dimensional Surface Plot

A three-dimensional surface plot is a projection of a three-dimensional surface such as a bivariate probability density function or other mathematical surface onto a plane. The shape of the surface can be displayed by using lines, or color and shading, or both. The following steps create a surface plot of a bivariate normal density function and display the results in Figure 1.94:

```
data normal;
   do x = -2 to 2 by 0.1;
      do y = -2 to 2 by 0.1;
         z = 0.164 * exp(-0.5 * ((x + y * 0.25) * x + (x * 0.25 + y) * y));
         output;
      end;
   end;
run;

proc template;
   define statgraph surfaceplotparm1;
      begingraph;
         entrytitle "Surface Plot of Normal Density Function";
         layout overlay3d;
            surfaceplotparm x=x y=y z=z;
         endlayout;
      endgraph;
   end;
run;
```

```
proc sgrender data=normal template=surfaceplotparm1;
run;
```

The LAYOUT OVERLAY3D statement is used when making three-dimensional plots such as a three-dimensional surface plot. The SURFACEPLOTPARM statement creates the surface from the X axis, Y axis, and Z axis variables specified on the X=, Y=, and Z= options. By default, the full cubic frame is drawn, and the surface is displayed by a fill color and grid lines. This corresponds to the default option SURFACETYPE=FILLGRID. Other options include SURFACETYPE=FILL for fill only and SURFACETYPE=WIREFRAME for a wire-frame grid only.

The following steps create a surface plot and display the results in Figure 1.95:

```
proc template;
   define statgraph surfaceplotparm2;
      begingraph;
         entrytitle "Surface Plot of Normal Density Function";
         layout overlay3d / cube=false;
            surfaceplotparm x=x y=y z=z / surfacetype=wireframe
                                          fillattrs=(color=black);
         endlayout;
      endgraph;
   end;
run;

proc sgrender data=normal template=surfaceplotparm2;
run;
```

The CUBE=FALSE option on the LAYOUT OVERLAY3D statement suppresses the display of the full cubic frame. The SURFACETYPE=WIREFRAME option suppresses the fill, and the FILLATTRS=(COLOR=BLACK) option sets the color of the wire-frame grid to black.

Figure 1.96 Three-Dimensional Histogram

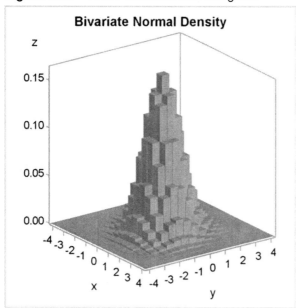

Figure 1.97 Three-Dimensional Histogram

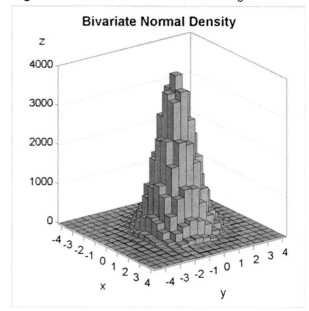

1.18   Three-Dimensional Histogram

A three-dimensional histogram displays tabulated crosstabulations, percents, density, and so on as solid bars. Each bar represents a nonoverlapping interval of two quantitative variables. The resulting three-dimensional representation is projected onto a plane for display. The following steps create a three-dimensional histogram of a bivariate normal density function and display the results in Figure 1.96:

```
data normal;
   do x = -4 to 4 by 0.5;
      do y = -4 to 4 by 0.5;
         z = 0.164 * exp(-0.5 * ((x + y * 0.25) * x + (x * 0.25 + y) * y));
         output;
      end;
   end;
run;

proc template;
   define statgraph bihistogram1;
      begingraph;
         entrytitle "Bivariate Normal Density";
         layout overlay3d / cube=false;
            bihistogram3dparm x=x y=y z=z;
         endlayout;
      endgraph;
   end;
run;

proc sgrender data=normal template=bihistogram1;
   run;
```

The LAYOUT OVERLAY3D statement is used when making three-dimensional plots. The CUBE=FALSE option on the LAYOUT OVERLAY3D statement suppresses the display of the full cubic frame. The BIHISTOGRAM3DPARM statement creates the three-dimensional histogram from the X axis, Y axis, and Z axis variables specified on the X=, Y=, and Z= options. Statements with a name that contains "PARM" do not do computations to summarize the data. They take results that are already summarized and are suitable for display. In this case, the data set provides a complete grid of X and Y values with a grid increment of 0.5 in both directions. The next example shows how to handle data that are not already in this form.

The following steps create a three-dimensional histogram of a random sample of observations from a bivariate normal distribution and display the results in Figure 1.97:

```
data sample(drop=i);
   do i = 1 to 100000;
      x = normal(104);
      y = -0.25 * x + sqrt(1 - 0.25 ** 2) * normal(104);
      output;
   end;
run;

data sample2;
   set sample;
   x = round(x, 0.5);
   y = round(y, 0.5);
run;

proc summary data=sample2 nway completetypes;
   class x y;
   var y;
   output out=sample3(keep=x y z) n=z;
run;

proc template;
   define statgraph bihistogram2;
      begingraph;
         entrytitle "Bivariate Normal Density";
         layout overlay3d / cube=false zaxisopts=(griddisplay=on);
            bihistogram3dparm x=x y=y z=z / display=all;
         endlayout;
      endgraph;
   end;
run;

proc sgrender data=sample3 template=bihistogram2;
run;
```

The first DATA step creates the random sample of observations. The second DATA step rounds them to the nearest multiple of 0.5. The PROC SUMMARY step counts how many values occur for every pair of rounded X and Y values. The NWAY option outputs only *n*-way (in this case two-way) frequencies. The COMPLETETYPES option creates all possible combinations of the classification variables even if the combination does not occur in the input data set. This option is required, because the BIHISTOGRAM3DPARM option requires a complete grid. The number of observations in each X and Y pair are stored in the variable Z that is specified in the N= option.

In the GTL, the `zAxisOpts=(GridDisplay=On)` option in the LAYOUT OVERLAY3D statement displays grid lines along the vertical or Z axis. The DISPLAY=ALL option in the BIHISTOGRAM3DPARM statement displays outlined and filled bins. By default, filled bins with no outlines are produced (as in Figure 1.96).

1.19 References

Cleveland, W. S., Devlin, S. J., and Grosse, E. (1988), "Regression by Local Fitting," *Journal of Econometrics*, 37, 87–114.

Eilers, P. H. C. and Marx, B. D. (1996), "Flexible Smoothing with *B*-Splines and Penalties," *Statistical Science*, 11, 89–121, with discussion.

Kuhfeld, W. F. (2009), "Modifying ODS Statistical Graphics Templates in SAS 9.2," `http://support.sas.com/rnd/app/papers/modtmplt.pdf`.

National Institute of Standards and Technology (1998), "Statistical Reference Data Sets," `http://www.nist.gov/srd/index.htm`, last accessed January 29, 2010.

SAS Institute Inc. (2008), *Base SAS 9.2 Procedures Guide: Statistical Procedures*, Cary, NC: SAS Institute Inc.

Chapter 2
Panels

Contents

Figure 2.1 Scatter Plot Matrix with Custom Template

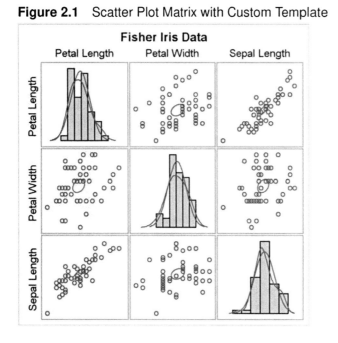

Figure 2.2 Scatter Plot Matrix with PROC SGSCATT

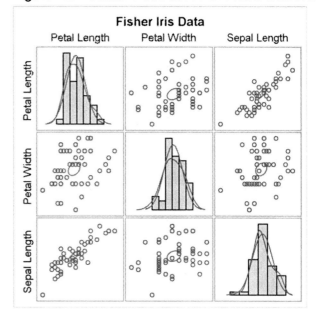

2.1   Scatter Plot Matrix

A scatterplot matrix is a rectangular display of scatter plots and potentially other types of graphs. It is common for a scatter plot matrix to provide all possible scatter plots for a list of variables along with univariate distributional information displayed on the diagonal of the matrix. The following steps create a scatter plot matrix using a custom template and PROC SGRENDER and also by using PROC SGSCATTER:

```
proc format;
    value specname   1='Setosa    '   2='Versicolor'   3='Virginica ';
run;

data iris;
    input SepalLength SepalWidth PetalLength PetalWidth Species @@;
    format Species specname.;
    label SepalLength='Sepal Length'     SepalWidth ='Sepal Width'
          PetalLength='Petal Length'     PetalWidth ='Petal Width';
    datalines;
50 33 14 02 1 64 28 56 22 3 65 28 46 15 2 67 31 56 24 3

    ... more lines ...

;
```

```
proc template;
   define statgraph matrix;
      begingraph / designheight=defaultdesignwidth;
         entrytitle 'Fisher Iris Data';
         layout gridded;
            scatterplotmatrix petallength petalwidth sepallength /
                     ellipse=(type=mean) diagonal=(histogram normal kernel);
         endlayout;
      EndGraph;
   end;
run;

proc sgrender data=iris(where=(species eq 3))
            template=matrix;
run;

proc sgscatter data=iris(where=(species eq 3));
   title 'Fisher Iris Data';
   matrix petallength petalwidth sepallength /
         ellipse=(type=mean) diagonal=(histogram normal kernel);
run;
```

The BEGINGRAPH statement has the option DESIGNHEIGHT=DEFAULTDESIGNWIDTH, which sets the height of the outer box that contains the graph to the default box width. The default design width is 640 pixels, and the default design height is 480 pixels. Hence, DESIGN-HEIGHT=DEFAULTDESIGNWIDTH and DESIGNHEIGHT=640px are equivalent. The plot can actually be produced at any size, but by default it is designed at a size that is $480/640 = 3/4$ as high as it is wide. The option DESIGNHEIGHT=DEFAULTDESIGNWIDTH is used to make a large square graph, designed for a size of 640 pixels by 640 pixels. Other common sizes in SAS/STAT templates include DESIGNWIDTH=DEFAULTDESIGNHEIGHT (a 480-pixel by 480-pixel square design) and DESIGNHEIGHT=360px (which works well for producing two side-by-side square plots).

All graph templates must contain a LAYOUT statement. Often, a LAYOUT OVERLAY statement or a LAYOUT LATTICE statement is used. However, this template uses a LAYOUT GRIDDED statement along with a SCATTERPLOTMATRIX statement. The SCATTERPLOTMATRIX statement, unlike all plotting statements discussed previously, cannot be specified in a LAYOUT OVERLAY block. The SCATTERPLOTMATRIX statement specifies the variables to be plotted. It also specifies options to produce a histogram on the diagonal along with normal and kernel density functions and a confidence ellipse for the means in the off diagonal plots. The results are displayed in Figure 2.1.

In the SG procedures, a scatter plot matrix is produced with PROC SGSCATTER and the MATRIX statement. Previous examples rely heavily on PROC SGPLOT. This is because previous examples produced single plots. You can use PROC SGSCATTER when you want to make a panel consisting of multiple plots. The remaining syntax is the same as the SCATTERPLOTMATRIX statement in the GTL. The results are displayed in Figure 2.2.

The following steps create additional plots; however, this time a 1 by 3 matrix and a 2 by 2 matrix of plots are created:

```
proc template;
   define statgraph compare1;
      begingraph;
         entrytitle 'Fisher Iris Data';
         layout gridded;
            scatterplotmatrix petalwidth sepallength sepalwidth /
                              rowvars=(petallength);
         endlayout;
      EndGraph;
   end;
run;

proc sgrender data=iris template=compare1;
run;

proc sgscatter data=iris;
   title 'Fisher Iris Data';
   compare y=petallength x=(petalwidth sepallength sepalwidth);
run;

proc template;
   define statgraph compare2;
      begingraph / designheight=defaultdesignwidth;
         entrytitle 'Fisher Iris Data';
         layout gridded;
            scatterplotmatrix sepallength sepalwidth /
                              rowvars=(petallength petalwidth );
         endlayout;
      EndGraph;
   end;
run;

proc sgrender data=iris template=compare2;
run;

proc sgscatter data=iris;
   title 'Fisher Iris Data';
   compare y=(petallength petalwidth) x=(sepallength sepalwidth);
run;
```

In the GTL, the ROWVARS= option specifies the variables that appear on the rows of the matrix, and the variables that appear on the columns are specified before the slash. In PROC SGSCATTER, the Y= option specifies the variables that appear on the rows of the matrix, and the X= option specifies the variables that appear on the columns of the matrix. The results are displayed in Figure 2.3, Figure 2.4, Figure 2.5, and Figure 2.6. In all four graphs, the component plots share axes for the variables that each pair of plots has in common.

The plots are not identical. The GTL uses a system where the ticks, tick labels, and axis labels alternate from side to side. In contrast, PROC SGSCATTER labels the axes consistently on the left and on the bottom. The differences between these plots are investigated next.

Figure 2.3 1 by 3 Matrix with Custom Template

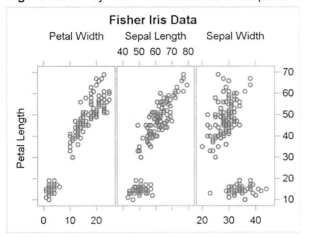

Figure 2.4 1 by 3 Matrix with PROC SGSCATTER

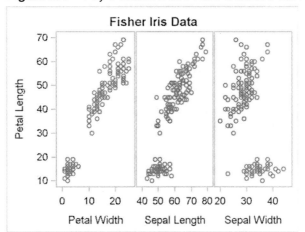

Figure 2.5 2 by 2 Matrix with Custom Template

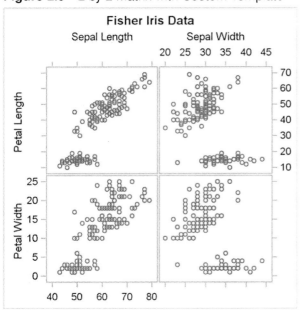

Figure 2.6 2 by 2 Matrix with PROC SGSCATTER

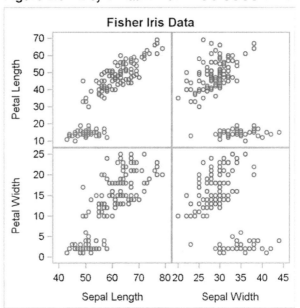

The SG procedures work by writing a template in the GTL and then by using it to produce a graph. PROC SGPLOT and PROC SGPANEL have an option on the PROC statement, the TMPLOUT= option, that writes the generated template to a file. You can look at that template and even use or modify it for use with PROC SGRENDER. The following step illustrates this option:

```
proc sgscatter data=iris tmplout='template.sas';
   title 'Fisher Iris Data';
   compare y=(petallength petalwidth) x=(sepallength sepalwidth);
run;
```

The generated template is displayed in Figure 2.7. Clearly, PROC SGSCATTER with the COMPARE statement takes a much different approach to creating these plots than does a SCATTERPLOTMA-TRIX statement in the GTL.

Figure 2.7 COMPARE Statement Generated Template

```
proc template;
define statgraph sgscatter;
begingraph / designwidth=640 designheight=640;
EntryTitle "Fisher Iris Data" /;
layout lattice / pad=(top=5) columns=2 columnDataRange=union rowDataRange=union;
   ColumnAxes;
   ColumnAxis / displaysecondary=none;
   ColumnAxis / displaysecondary=none;
   EndColumnAxes;
   RowAxes;
   RowAxis / displaysecondary=none;
   RowAxis / displaysecondary=none;
   EndRowAxes;
   ScatterPlot X=SepalLength Y=PetalLength / primary=true;
   ScatterPlot X=SepalWidth Y=PetalLength / primary=true;
   ScatterPlot X=SepalLength Y=PetalWidth / primary=true;
   ScatterPlot X=SepalWidth Y=PetalWidth / primary=true;
endlayout;
endgraph;
end;
run;
```

If you want to make a graph, and if you know how to make something similar to what you want with one of the SG procedures, then you can use the SG procedure to generate a template and use that as a starting point. You can then modify it to get precisely what you want. This might be easier than writing a template from scratch.

The following steps produce a panel of independent plots, each with separate and independent axes:

```
proc template;
   define statgraph scatter;
      begingraph / designheight=defaultdesignwidth;
         entrytitle "Fisher Iris Data";
         layout lattice / rows=2 columns=2 rowgutter=10 columngutter=10;
            scatterplot x=petalwidth y=petallength;
            scatterplot x=sepalwidth y=sepallength;
            scatterplot x=sepallength y=petallength;
            scatterplot x=sepalwidth y=petalwidth;
         endlayout;
      endgraph;
   end;
run;


proc sgrender data=iris template=scatter;
run;
```

Figure 2.8 Panel of Plots with Custom Template

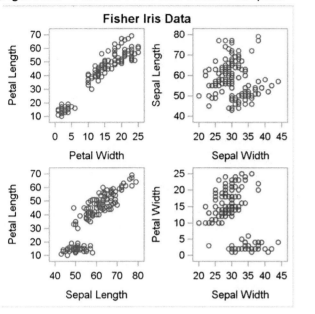

Figure 2.9 Panel of Plots with PROC SGSCATTER

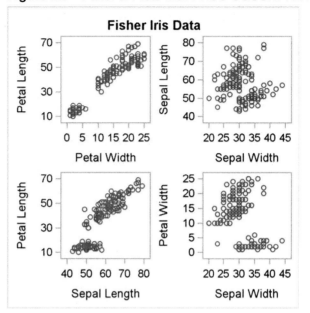

```
ods graphics on / height=640px width=640px;

proc sgscatter data=iris;
   title 'Fisher Iris Data';
   plot petallength * petalwidth sepallength * sepalwidth
        petallength * sepallength petalwidth * sepalwidth;
run;

ods graphics on / reset=all;
```

In the GTL, the LAYOUT LATTICE statement is used to create a panel of multiple plots. This particular panel has two rows, two columns, and four SCATTERPLOT statements. By default, there is no space between the rows and columns. In this template, the option ROWGUTTER=10 inserts 10 pixels between rows, and the option COLUMNGUTTER=10 inserts 10 pixels between columns. Note, however, that this is 10 pixels relative to the designed size of the plot (480px by 640px by default and 640px by 640px here due to the design height being set to the default design width). If the plot is shrunk, then the 10 pixels are shrunk accordingly. The four SCATTERPLOT statements create the four plots. The results are displayed in Figure 2.8.

In PROC SGSCATTER, the PLOT statement contains a list of each pair of variables to plot, with an asterisk between each pair. The ODS GRAPHICS statement specifies HEIGHT=640PX so that the plot size will match the plot in Figure 2.8. The final ODS GRAPHICS statement restores the default height. The results are displayed in Figure 2.9. The two panels are identical.

Figure 2.10   Panel of Plots with Custom Template

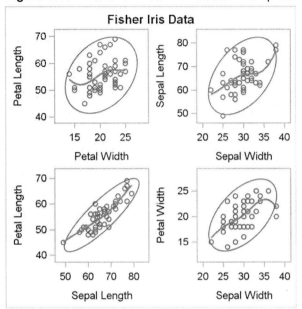

Figure 2.11   Panel of Plots with PROC SGSCATTE

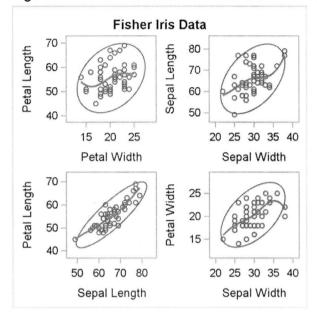

The following template is equivalent to the template that was used to make Figure 2.8:

```
proc template;
   define statgraph scatter;
      begingraph;
         entrytitle "Fisher Iris Data";
         layout lattice / rows=2 columns=2 rowgutter=10 columngutter=10;
            layout overlay;
               scatterplot x=petalwidth y=petallength;
            endlayout;
            layout overlay;
               scatterplot x=sepalwidth y=sepallength;
            endlayout;
            layout overlay;
               scatterplot x=sepallength y=petallength;
            endlayout;
            layout overlay;
               scatterplot x=sepalwidth y=petalwidth;
            endlayout;
         endlayout;
      endgraph;
   end;
run;
```

If you want to have multiple plotting statements in each graph, then you need to have a LAYOUT OVERLAY statement that groups them. The following steps illustrate:

```
proc template;
   define statgraph scatter;
      begingraph / designheight=defaultdesignwidth;
         entrytitle "Fisher Iris Data";
         layout lattice / rows=2 columns=2 rowgutter=10 columngutter=10;
            layout overlay;
               scatterplot x=petalwidth y=petallength;
               ellipse x=petalwidth y=petallength / type=predicted;
               pbsplineplot x=petalwidth y=petallength;
            endlayout;
            layout overlay;
               scatterplot x=sepalwidth y=sepallength;
               ellipse x=sepalwidth y=sepallength / type=predicted;
               pbsplineplot x=sepalwidth y=sepallength;
            endlayout;
            layout overlay;
               scatterplot x=sepallength y=petallength;
               ellipse x=sepallength y=petallength / type=predicted;
               pbsplineplot x=sepallength y=petallength;
            endlayout;
            layout overlay;
               scatterplot x=sepalwidth y=petalwidth;
               ellipse x=sepalwidth y=petalwidth / type=predicted;
               pbsplineplot x=sepalwidth y=petalwidth;
            endlayout;
         endlayout;
      endgraph;
   end;
run;

proc sgrender data=iris(where=(species=3)) template=scatter;
run;

ods graphics on / height=640px width=640px;

proc sgscatter data=iris(where=(species=3));
   title 'Fisher Iris Data';
   plot petallength * petalwidth sepallength * sepalwidth
        petallength * sepallength petalwidth * sepalwidth /
        ellipse=(type=predicted) pbspline;
run;

ods graphics on / reset=all;
```

Each LAYOUT OVERLAY block contains three plotting statements. The PROC SGSCATTER step simply adds the following options to make the same plot: `ellipse=(type=predicted) pbspline`. Options from many of the statements that are discussed in previous examples are available on the PLOT and COMPARE statements. The results are displayed in Figure 2.10 and Figure 2.11.

You can simplify writing templates like these by using the macro language, for example, as follows:

```
proc template;
   define statgraph scatter4;
      begingraph / designheight=defaultdesignwidth;
         entrytitle "Fisher Iris Data";
         layout lattice / rows=2 columns=2 rowgutter=10 columngutter=10;

            %macro layout(v1,v2);
               layout overlay;
                  scatterplot  x=&v1 y=&v2;
                  ellipse      x=&v1 y=&v2 / type=predicted;
                  pbsplineplot x=&v1 y=&v2;
               endlayout;
            %mend;

            %layout(petalwidth,  petallength)

            %layout(sepalwidth,  sepallength)

            %layout(sepallength, petallength)

            %layout(sepalwidth,  petalwidth)
         endlayout;
      endgraph;
   end;
run;
```

This step generates the same template that was used to make Figure 2.10.

Figure 2.12 Basic Residual Panel

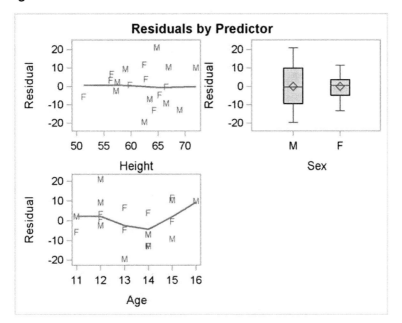

2.2 Residual Panel

This example uses PROC GLM to fit a linear model and then uses PROC TEMPLATE, the GTL, and PROC SGRENDER to display the residuals as a function of each of the predictor variables. This example introduces the CELL, ENDCELL, CELLHEADER, ENDCELLHEADER, SIDEBAR, and ENDSIDEBAR statements and shows you how to construct common axes and axis labels. The following step fits a linear model with PROC GLM and creates an output data set called Res that contains all of the analysis variables along with a new variable, r, which contains the residuals:

```
proc glm data=sashelp.class;
   class sex;
   model weight = height age sex;
   output out=res r=r;
run;
```

The linear model has two quantitative and continuous predictor variables, Height and Age, and one categorical predictor variable, Sex. Residuals are displayed in scatter plots for the continuous predictor variables and in box plots for the categorical predictor variable.

The following step creates a template in the form of a two by two lattice, containing three residual plots, one for each of the predictor variables:

```
proc template;
   define statgraph res;
      begingraph;
         entrytitle 'Residuals by Predictor';

         layout lattice / rows=2 columns=2;

            layout overlay;
               scatterplot y=r x=height / markercharacter=sex group=sex;
               loessplot   y=r x=height;
            endlayout;

            layout overlay;
               boxplot y=r x=sex;
            endlayout;

            layout overlay;
               scatterplot y=r x=age / markercharacter=sex group=sex;
               loessplot   y=r x=age;
            endlayout;

         endlayout;
      endgraph;
   end;
run;
```

This template is a starting point and is refined several times to illustrate various statements, options, and functionality. Inside the LAYOUT LATTICE block are three LAYOUT OVERLAY blocks. The first LAYOUT OVERLAY block creates a scatter plot of the residuals as a function of the variable Height. Each observation is displayed as a red 'F' for females or as a blue 'M' for males. The symbol comes from the MARKERCHARACTER=SEX option, and the color comes from the GROUP=SEX option. Colors are assigned in data order, not alphabetical order ('M' comes before 'F' in the data set). A loess fit plot is overlaid on the scatter plot to see if there is any systematic pattern in the residuals. The second LAYOUT OVERLAY block creates a box plot of the residuals for males and one for females. The third LAYOUT OVERLAY block is like the first, but it is for the variable Age. The following step creates the plot, which is displayed in Figure 2.12:

```
proc sgrender data=res template=res;
label r ='Residual';
run;
```

Figure 2.13 Panel with Titles and Labels

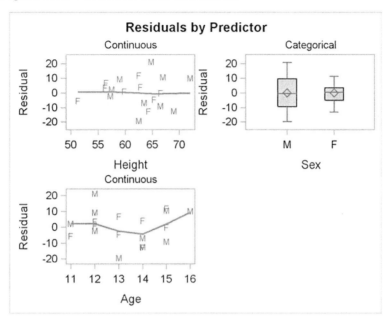

The following PROC TEMPLATE step is similar to the previous one, except now each graph has a title and each Y axis has a label explicitly set in the template:

```
proc template;
   define statgraph res;
      begingraph;
         entrytitle 'Residuals by Predictor';
         layout lattice / rows=2 columns=2;

            layout overlay / yaxisopts=(label='Residual');
               entry "Continuous" / location=outside;
               scatterplot y=r x=height / markercharacter=sex group=sex;
               loessplot   y=r x=height;
            endlayout;

            layout overlay / yaxisopts=(label='Residual');
               entry "Categorical" / location=outside;
               boxplot y=r x=sex;
            endlayout;

            layout overlay / yaxisopts=(label='Residual');
               entry "Continuous" / location=outside;
               scatterplot y=r x=age / markercharacter=sex group=sex;
               loessplot   y=r x=age;
            endlayout;

         endlayout;
      endgraph;
   end;
run;
```

An ENTRY statement inside the LAYOUT OVERLAY block with the option LOCATION=OUTSIDE creates a header or title for each graph. The option `yAxisOpts=(Label='Residual')` in each LAYOUT OVERLAY statement creates the label. The following step creates the plot, which is displayed in Figure 2.13:

```
proc sgrender data=res template=res;
run;
```

The previous PROC TEMPLATE step created a title for each graph by using the ENTRY statement. If you need more than one title, the approach illustrated in the following step works better:

```
proc template;
   define statgraph res;
      begingraph;
         layout lattice / rows=2 columns=2;

            cell;
               cellheader;
                  entry "Continuous" / textattrs=(weight=bold);
                  entry "Predictor" / textattrs=(weight=bold);
               endcellheader;
               layout overlay / yaxisopts=(label='Residual');
                  scatterplot y=r x=height / markercharacter=sex group=sex;
                  loessplot   y=r x=height;
               endlayout;
            endcell;

            cell;
               cellheader;
                  entry "Categorical" / textattrs=(weight=bold);
                  entry "Predictor" / textattrs=(weight=bold);
               endcellheader;
               layout overlay / yaxisopts=(label='Residual');
                  boxplot y=r x=sex;
               endlayout;
            endcell;

            cell;
               cellheader;
                  entry "Continuous" / textattrs=(weight=bold);
                  entry "Predictor" / textattrs=(weight=bold);
               endcellheader;
               layout overlay / yaxisopts=(label='Residual');
                  scatterplot y=r x=age / markercharacter=sex group=sex;
                  loessplot   y=r x=age;
               endlayout;
            endcell;

         endlayout;
      endgraph;
   end;
run;
```

Figure 2.14 Panel Using Cells

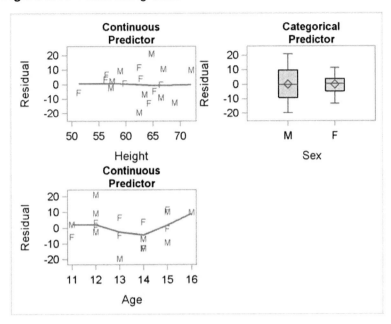

In this template, each LAYOUT OVERLAY block is wrapped in a CELL statement block beginning with a CELL statement and ending with an ENDCELL statement. Inside the CELL statement block is a CELLHEADER statement block beginning with a CELLHEADER statement and ending with an ENDCELLHEADER statement. Inside of that you can provide one or more titles, each in an ENTRY statement. You could use multiple ENTRY statements with LOCATION=OUTSIDE and without the CELL blocks; however, the spacing between titles is nicer with this approach. The following step creates the plot, which is displayed in Figure 2.14:

```
proc sgrender data=res template=res;
run;
```

The following PROC TEMPLATE step creates a template for the residual panel, but this time inserts additional space between the plots:

```
proc template;
   define statgraph res;
      begingraph;
         entrytitle 'Residuals by Predictor';
         layout lattice / rows=2 columns=2 rowgutter=20 columngutter=100;

            layout overlay / yaxisopts=(label='Residual');
               entry "Continuous" / location=outside;
               scatterplot y=r x=height / markercharacter=sex group=sex;
               loessplot    y=r x=height;
            endlayout;

            layout overlay / yaxisopts=(label='Residual');
               entry "Categorical" / location=outside;
               boxplot y=r x=sex;
            endlayout;
```

Figure 2.15 Panel with Gutter Space

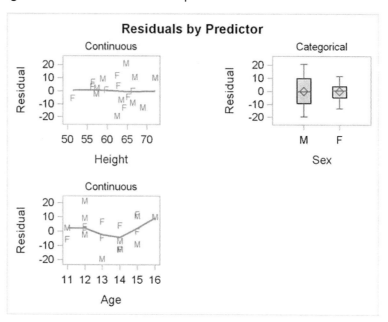

```
      layout overlay / yaxisopts=(label='Residual');
         entry "Continuous" / location=outside;
         scatterplot y=r x=age / markercharacter=sex group=sex;
         loessplot    y=r x=age;
      endlayout;
   endlayout;
endgraph;
end;
run;
```

The plots in Figure 2.12 through Figure 2.14 are all close to each of the other plots in their panel. This template uses the options `RowGutter=20 ColumnGutter=100` on the LAYOUT LATTICE statement to insert 20 pixels of white space between each row of plots and 100 pixels of white space between each column. The following step creates the plot, which is displayed in Figure 2.15:

```
proc sgrender data=res template=res;
run;
```

The following steps add several new features including a common axis label:

```
proc template;
   define statgraph res;
      begingraph;
         entrytitle 'Residuals by Predictor';
         layout lattice / rows=2 columns=2 rowgutter=20 columngutter=20;

            layout overlay / xaxisopts=(offsetmin=0.1 offsetmax=0.1)
                             yaxisopts=(offsetmin=0.1 offsetmax=0.1
                             display=(line ticks tickvalues));
               scatterplot y=r x=height / markercharacter=sex group=sex;
               loessplot   y=r x=height;
            endlayout;

            layout overlay / yaxisopts=(display=(line ticks tickvalues));
               boxplot y=r x=sex;
            endlayout;

            layout overlay / xaxisopts=(offsetmin=0.1 offsetmax=0.1)
                             yaxisopts=(offsetmin=0.1 offsetmax=0.1
                             display=(line ticks tickvalues));
               scatterplot y=r x=age / markercharacter=sex group=sex;
               loessplot   y=r x=age;
            endlayout;

            sidebar / align=left;
               entry "Residual" / rotate=90;
            endsidebar;

            sidebar / align=bottom;
               entry "Analysis of the Class Data Set - July 25, 2009" /
                     textattrs=(weight=bold);
            endsidebar;

         endlayout;
      endgraph;
   end;
run;
```

Each LAYOUT OVERLAY statement had the option `yAxisOpts=(display=(line ticks TickValues))` which differs from the default in that the axis label is omitted. Instead of individual axis labels, a single label is placed in the far left or "side bar" area by the following statements:

```
sidebar / align=left;
   entry "Residual" / rotate=90;
endsidebar;
```

Figure 2.16 Common Axis Label

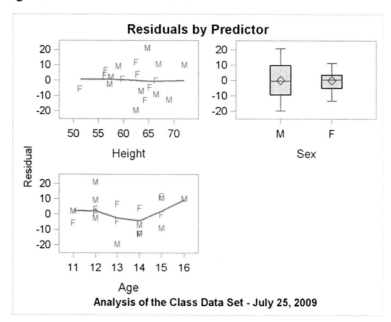

The label "Residual" is printed on the left and rotated 90 degrees to print vertically. An additional side bar with ALIGN=BOTTOM is used to produce a note at the bottom of the display, although this could have been done more easily with a ENTRYFOOTNOTE statement. Side bar information can be placed in the top, bottom, left, and right of the graph. Additional options in this template include ROWGUTTER=20 and COLUMNGUTTER=20 in the LAYOUT LATTICE statement to place 20 pixels of white space between each graph. The options **xAxisOpts=(OffsetMin=0.1 OffsetMax=0.1) yAxisOpts=(OffsetMin=0.1 OffsetMax=0.1)** add white space (roughly 10%) between the extreme data values and the axes, which for some plots improves the appearance of the display.

The following step creates the plot, which is displayed in Figure 2.16:

```
proc sgrender data=res template=res;
run;
```

The following PROC TEMPLATE step replaces the common axis label in the side bar with a common label for each row of the plot and uses the tick marks and tick labels from the plot on the left for the plot on the right as well:

```
proc template;
   define statgraph res;
      begingraph;
         entrytitle 'Residuals by Predictor';
         layout lattice / rows=2 columns=2 rowgutter=20
                          rowdatarange=unionall;
         rowaxes;
            rowaxis / label="Residual";
            rowaxis / label="Residual";
         endrowaxes;
```

Figure 2.17 Common Axes

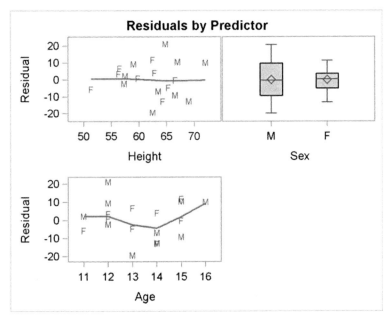

```
layout overlay / xaxisopts=(offsetmin=0.1 offsetmax=0.1)
                 yaxisopts=(offsetmin=0.1 offsetmax=0.1);
    scatterplot y=r x=height / markercharacter=sex group=sex;
    loessplot   y=r x=height;
endlayout;

layout overlay;
    boxplot y=r x=sex;
endlayout;

layout overlay / xaxisopts=(offsetmin=0.1 offsetmax=0.1)
                 yaxisopts=(offsetmin=0.1 offsetmax=0.1);
    scatterplot y=r x=age / markercharacter=sex group=sex;
    loessplot   y=r x=age;
endlayout;
            endlayout;
          endgraph;
        end;
    run;
```

The LAYOUT LATTICE statement has the option ROWDATARANGE=UNIONALL which creates
a common Y axis. The following statements create the labels for the common axes:

```
rowaxes;
    rowaxis / label="Residual";
    rowaxis / label="Residual";
endrowaxes;
```

There is one ROWAXIS statement for each row. Similarly, there are COLUMNAXES, COLUM-
NAXIS, and ENDCOLUMNAXES statements, although they are not shown here.

Figure 2.18 Lattice Layout

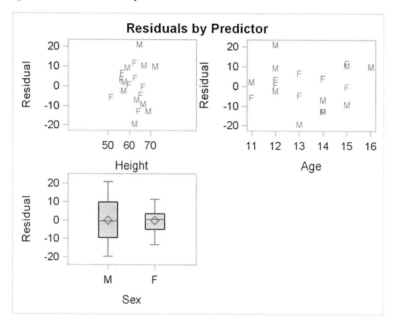

The following step creates the plot, which is displayed in Figure 2.17:

```
proc sgrender data=res template=res;
run;
```

The last part of this example illustrates a difference between the lattice layout and the gridded layout. The following two templates are similar to the template used to make Figure 2.12, but the loess fits and all extra options and statements have been removed. Also, the box plot has been moved from the first row to the second row. The following two templates differ only in that the first uses a LAYOUT LATTICE statement, like before, and the second instead uses a LAYOUT GRIDDED statement:

```
proc template;
   define statgraph res1;
      begingraph;
         entrytitle 'Residuals by Predictor';
         layout lattice / rows=2 columns=2;
            scatterplot y=r x=height / markercharacter=sex group=sex;
            scatterplot y=r x=age / markercharacter=sex group=sex;
            boxplot y=r x=sex;
         endlayout;
      endgraph;
   end;
```

Figure 2.19 Gridded Layout

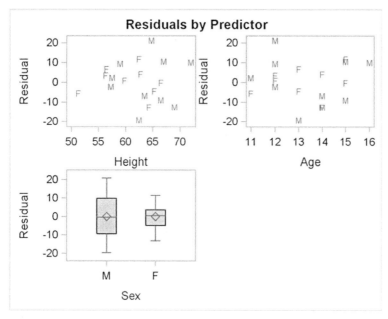

```
define statgraph res2;
   begingraph;
      entrytitle 'Residuals by Predictor';
      layout gridded / rows=2 columns=2;
         scatterplot y=r x=height / markercharacter=sex group=sex;
         scatterplot y=r x=age / markercharacter=sex group=sex;
         boxplot y=r x=sex;
      endlayout;
   endgraph;
end;
run;
```

The following steps use these templates to create Figure 2.18 and Figure 2.19:

```
proc sgrender data=res template=res1;
   label r = 'Residual';
run;

proc sgrender data=res template=res2;
   label r = 'Residual';
run;
```

The lattice layout aligns the plot and tick display areas across graphs to facilitate comparisons. The box plot in Figure 2.18 and the scatter plot of residuals by height have a common tick display area. In contrast, with the gridded layout in Figure 2.19, the axes are independent. In many cases, the alignment that the lattice layout provides is preferable. In some cases it is not, and you should use the gridded layout instead.

Figure 2.20 Data Panel with Custom Template

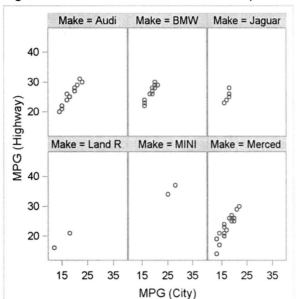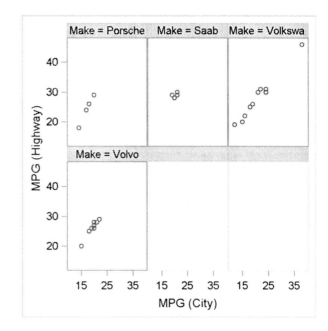

2.3 Data Panel

A data panel is a matrix of graphs, with one graph for each level of a classification variable (or for each combination of levels of two or more classification variables). The following steps produce a data panel of scatter plots by using the GTL and PROC SGRENDER and also by using PROC SGPANEL:

```
proc template;
   define statgraph panel;
      begingraph;
         layout datapanel classvars=(make) / rows=2 columns=3;
            layout prototype;
               scatterplot x=mpg_city y=mpg_highway;
            endlayout;
         endlayout;
      endgraph;
   end;
run;

proc sgrender data=sashelp.cars(where=(origin='Europe')) template=panel;
run;

proc sgpanel data=sashelp.cars(where=(origin='Europe'));
   title 'Cars by Make';
   panelby make / rows=2 columns=3;
   scatter x=mpg_city y=mpg_highway;
run;
```

The results are displayed in Figure 2.20 and Figure 2.21. The classification variable Make has 10 levels for this subset of the data set. Data panels with six plots (two rows and three columns) are

Figure 2.21   Data Panel with PROC SGPANEL

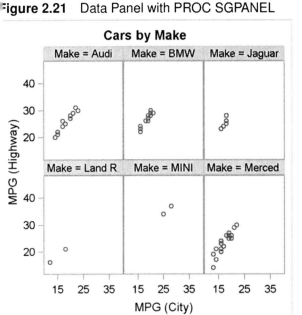
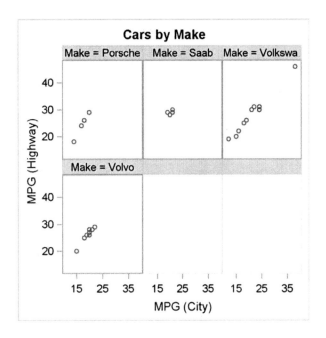

requested, so each step produces two panels, one with the first six graphs, and one with the remaining four graphs. Each graph has the same axes with the same ticks as every other graph both within and across panels. Within a panel, plots in the same row share a common Y axis and plots in the same column share a common X axis. The levels of the classification variable appear as headers over each graph.

In the GTL, the data panel is specified with a LAYOUT DATAPANEL statement, one or more classification variables specified in the CLASSVARS= option, and the number of rows and columns specified in the ROWS= and COLUMNS= options. You must specify a single LAYOUT PROTO-TYPE block within the LAYOUT DATAPANEL block. One or more plotting statements (in this case the SCATTERPLOT statement) is specified inside the LAYOUT PROTOTYPE block. These statements define the common graph elements to be placed inside of each cell of the panel. Each graph is constructed in the same way, but for a different level of the classification variable.

In contrast, a LAYOUT LATTICE statement with two rows and columns was used in the section "Residual Panel" on page 95. With the lattice layout, four plot specifications are required since the template produces a panel with four graphs. Also, with the data panel layout, one Y axis variable and one X axis variable are plotted for different levels of a classification variable. By contrast, in the lattice layout, a different variable can potentially appear on each axis of each plot, there is no classification variable, and each graph in the panel can be of a different type from every other graph.

PROC SGPANEL is the SG procedure that is used to make the data panels. The PANELBY statement specifies the classification variable Make, and the number of rows and columns. The SCATTER statement produces the scatter plot.

In Figure 2.20, which was created by using the GTL and PROC SGRENDER, the graphs appear in alphabetical order, sorted by the levels of the classification variable Make. This is because that is the order in which the levels appeared in the data set. In Figure 2.21, which was created by using PROC SGPANEL, the graphs appear in alphabetical or sorted order, because that is what PROC SGPANEL does. If you want any order other than the sorted order, you must use the GTL and PROC SGRENDER and put the data in the order that you want. You do not have to sort the entire data set in that order; you just need the first few observations to be in the right order. One way to control the order is to create new observations in the front of the data set with missing values in all of the variables except the classification variable. Create one observation for each level and insert them in the order that you want the graphs to appear. For example, you could do the following to sort by descending size:

```
data cars;
   if _n_ = 1 then do;
      type = 'Truck   '; output;
      type = 'SUV     '; output;
      type = 'Wagon   '; output;
      type = 'Sedan   '; output;
      type = 'Sports  '; output;
      type = 'Hybrid  '; output;
   end;
   set sashelp.cars;
   output;
run;
```

The next three examples use the same data set and two classification variables. The first example uses PROC SGPANEL in a way analogous to before and produces Figure 2.22. The second example uses LAYOUT DATAPANEL statement in a way analogous to before and produces Figure 2.23. An alternative layout for data such as these is the LAYOUT DATALATTICE statement. The third example uses a LAYOUT DATALATTICE statement and produces Figure 2.24.

Before beginning the next example, here are some examples of the LAYOUT DATAPANEL and LAYOUT DATALATTICE statements:

```
layout datapanel classvars=(cylinders type) / rows=2 columns=3;
layout datalattice columnvar=type rowvar=cylinders / rows=2 columns=3;
```

They differ in that the LAYOUT DATAPANEL statement does not assign specific row or column roles to the classification variables. By contrast, in the LAYOUT DATALATTICE statement, the variable Type variable is specified on the COLUMNVAR= option and the variable Cylinders is specified on the ROWVAR= option. The plots in Figure 2.23 and Figure 2.24 illustrate the graphical displays that the two different layouts produce.

The following step creates a data panel display, again by using PROC SGPANEL, and creates Figure 2.22:

```
proc sgpanel data=sashelp.cars(where=(cylinders in (4, 6)));
   title 'Cars by Cylinders and Type';
   panelby cylinders type / rows=2 columns=3 sparse;
   scatter x=mpg_city y=mpg_highway;
run;
```

Figure 2.22 Data Panel with PROC SGPANEL

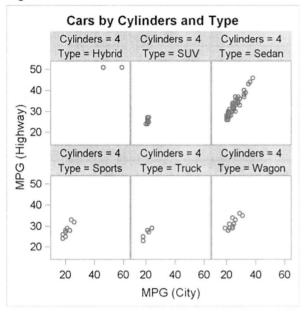
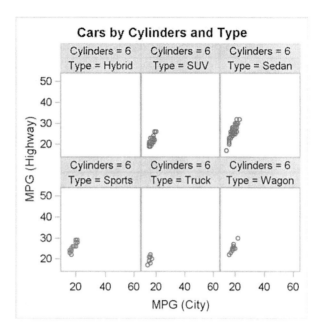

The following steps use the LAYOUT DATAPANEL statement and PROC SGRENDER and create Figure 2.23:

```
proc template;
   define statgraph panel;
      begingraph;
         layout datapanel classvars=(cylinders type) /
                           rows=2 columns=3 sparse=true;
            layout prototype;
               scatterplot x=mpg_city y=mpg_highway;
            endlayout;
         endlayout;
      endgraph;
   end;
run;

proc sgrender data=sashelp.cars(where=(cylinders in (4, 6)))
              template=panel;
run;
```

The syntax for the preceding SGPANEL and LAYOUT DATAPANEL statement examples is very similar to the previous data panel examples. Now, however, there are two classification variables in each list instead of one. The only other difference is the addition of the SPARSE=TRUE option to the LAYOUT DATAPANEL statement in the template and the SPARSE option to the PANELBY statement of PROC SGPANEL. These options enable the procedures to create empty cells for crossings of the classification variables that are not present in the input data set. By default, empty cells are not created. The effect of the SPARSE options is shown by the "Cylinders=6 Type=Hybrid" empty plot.

Figure 2.23 Data Panel with LAYOUT DATAPANEL

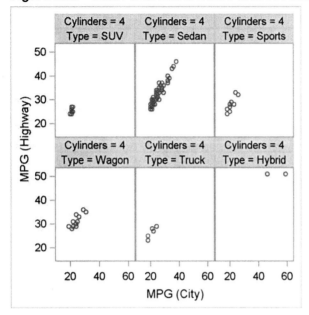

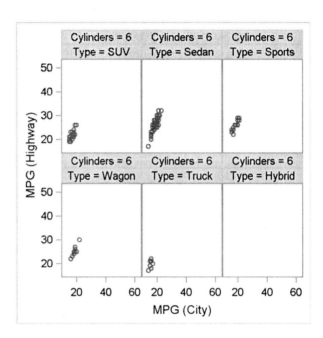

Figure 2.24 Data Panel with LAYOUT DATALATTICE

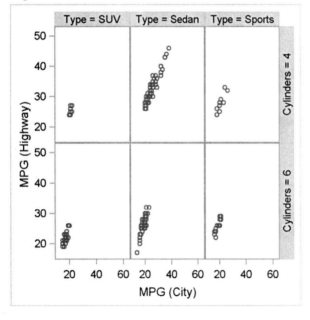

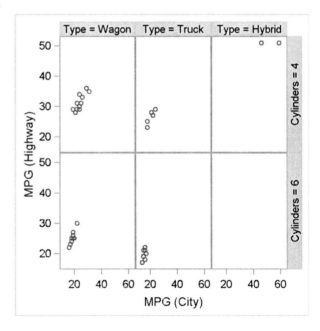

The following steps create plots for the combinations of the same two class variables, but this time using the LAYOUT DATALATTICE statement in the GTL:

```
proc template;
   define statgraph lattice;
      begingraph;
         layout datalattice columnvar=type rowvar=cylinders /
                            rows=2 columns=3;
            layout prototype;
               scatterplot x=mpg_city y=mpg_highway;
            endlayout;
         endlayout;
      endgraph;
   end;
run;


proc sgrender data=sashelp.cars(where=(cylinders in (4, 6)))
             template=lattice;
run;
```

The results are displayed in Figure 2.24. Figure 2.23 displays results using the LAYOUT DATA-PANEL statement, and Figure 2.24 displays results using the LAYOUT DATALATTICE statement. In Figure 2.23, each graph has a header that displays the names and levels of the two classification variables. Recall that the LAYOUT DATAPANEL statement syntax for specifying the classification variables is: `ClassVars=(cylinders type)`. In Figure 2.24, each graph has a header that displays the name of just the variable Type and its level, and a row header that displays the name of the variable Cylinders and its level. Recall that the LAYOUT DATALATTICE statement syntax for specifying the classification variables is: `ColumnVar=type RowVar=cylinders`. All of the graphs have the same axes.

When you have one classification variable and you want one row of panels in your graphical display, you can use the LAYOUT DATALATTICE statement. When you have one classification variable and you want more than one row of panels in your graphical display (such as in Figure 2.20), you cannot use the LAYOUT DATALATTICE statement and must use PROC SGPANEL or the LAYOUT DATAPANEL statement instead. When you have two classification variables, and you want the row and column header effect that is displayed in Figure 2.24, use the LAYOUT DATALATTICE statement. For most other uses you will want to use PROC SGPANEL or the LAYOUT DATAPANEL statement instead.

You can use PROC SGPANEL to get the same row and column headers as you get with the LAYOUT DATALATTICE statement by using the LAYOUT=LATTICE option in the PANELBY statement as follows:

```
proc sgpanel data=sashelp.cars(where=(cylinders in (4, 6)));
   title 'Cars by Type and Cylinders';
   panelby type cylinders / rows=2 columns=3 sparse layout=lattice;
   scatter x=mpg_city y=mpg_highway;
run;
```

The results of this step are not shown.

Figure 2.25 Data Panel with Custom Template

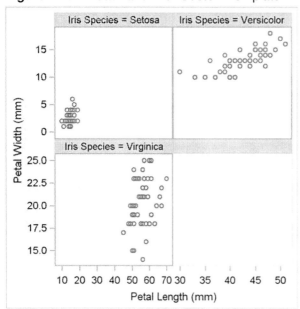

Figure 2.26 Data Panel with PROC SGPANEL

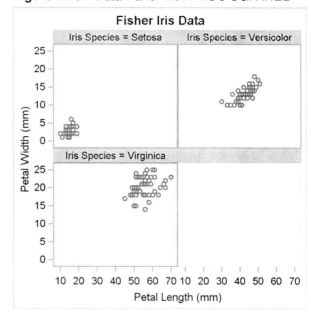

The following steps create a scatter plot of the iris data for each of the three species:

```
proc template;
   define statgraph panel;
      begingraph / designheight=defaultdesignwidth;
         layout datapanel classvars=(species) / rows=2 columns=2
                           rowdatarange=union columndatarange=union;
            layout prototype;
               scatterplot x=petallength y=petalwidth;
            endlayout;
         endlayout;
      endgraph;
   end;
run;


proc sgpanel data=sashelp.iris;
   title 'Fisher Iris Data';
   panelby species;
   scatter x=petallength y=petalwidth;
run;


proc sgrender data=sashelp.iris template=panel;
run;
```

The results are displayed in Figure 2.25 and Figure 2.26. In the template, the options `RowDataRange=union ColumnDataRange=union` are specified to enable ODS Graphics to create separate axis scales for each row and each column. The option in the PANELBY statement that controls row and column axis scaling is the UNISCALE= option, and it is not used in this example. The differences between Figure 2.25 and Figure 2.26 are due to the specification of the scaling options.

You should also note that as of SAS 9.2, computed plot statements (for example, BOXPLOT, DENSITYPLOT, ELLIPSE, HISTOGRAM, LOESSPLOT, MODELBAND, PBSPLINEPLOT, and REGRESSIONPLOT) cannot be used inside of the LAYOUT PROTOTYPE block which is inside the LAYOUT DATAPANEL and the LAYOUT DATALATTICE blocks. This might change in a future SAS release. The equivalent statements are, however, allowed with PROC SGPANEL.

Chapter 3

Style Templates

Contents

ODS styles control the colors and general appearance of all graphs and tables. Styles are composed of style elements that control specific aspects of graphs (for example, the font of a title, the color of a reference line, the style of a regression fit line, and so on). Graphs are constructed from three sources: a data object (in the case of an analytical procedure) or a SAS data set (in the case of PROC SGRENDER or an SG procedure), a graph template, and a style template. The following steps create a scatterplot from a graph template called `ClassScatter`, a style called STATISTICAL (which has a style template called `Styles.Statistical`), and a SAS data set called Sashelp.Class:

```
proc template;
   define statgraph classscatter;
      begingraph;
         entrytitle 'Weight By Height';
         layout overlay;
            scatterplot y=weight x=height / markerattrs=GraphData2;
         endlayout;
      endgraph;
   end;
run;

ods listing style=statistical;
proc sgrender data=sashelp.class template=classscatter;
run;
```

The results are not displayed, but the appearance of the markers in the scatterplot is controlled by the `GraphData2` style element in the STATISTICAL style.

This chapter describes the components of styles and some of the differences between them. The section "ODS Styles" on page 116 explains some of the basics of styles and style templates. The section "ODS Style Modification" on page 126 shows PROC TEMPLATE steps that you can use to modify individual style elements. The section "Colors and Groups" on page 140 explains colors, the

`GraphData`*n* style elements, and how styles are used to control the display of groups of observations (for example, with data grouped by sex or other categorical variables). The section "An All-Color Style" on page 145 explains how to construct a style where groups are only distinguished by color (and not by marker or line differences). The section "Color Changes in a Survival Plot" on page 148 demonstrates how to modify colors in the Kaplan-Meier survival plot. The section "Direct Style Modifications" on page 149 explains more about colors and provides additional examples of style modification.

3.1 ODS Styles

The definition of a style is provided by a style template. The following step lists the names of all style templates:

```
proc template;
   list styles;
run;
```

The results of this step are not displayed, but a few of the style templates in the list include: `Styles.Analysis`, `Styles.Default`, `Styles.Journal`, `Styles.Journal2`, `Styles.Listing`, `Styles.Rtf`, and `Styles.Statistical`. These templates define the styles that are most often used for statistical work. The STATISTICAL style is used by default in SAS/STAT documentation and in this book.

The following step displays the STATISTICAL style template:

```
proc template;
   source styles.statistical;
run;
```

The first two lines of the results are as follows:

```
define style Styles.Statistical;
   parent = styles.default;
```

The parent for the STATISTICAL style is the DEFAULT style, so the STATISTICAL style inherits many of its properties from the DEFAULT style. The following step displays the DEFAULT style:

```
proc template;
   source styles.default;
run;
```

The results do not contain a PARENT= option, so the STATISTICAL style has only one parent, namely the DEFAULT style. The STATISTICAL style is defined first by the DEFAULT style, but some aspects of the DEFAULT style are overridden. The first part of the STATISTICAL style that pertains to graphics is the following declaration of fonts:

```
style GraphFonts /
   'GraphDataFont' = ("<sans-serif>, <MTsans-serif>",7pt)
   'GraphUnicodeFont' = ("<MTsans-serif-unicode>",9pt)
   'GraphValueFont' = ("<sans-serif>, <MTsans-serif>",9pt)
   'GraphLabelFont' = ("<sans-serif>, <MTsans-serif>",10pt)
   'GraphFootnoteFont' = ("<sans-serif>, <MTsans-serif>",10pt,italic)
   'GraphTitleFont' = ("<sans-serif>, <MTsans-serif>",11pt,bold)
   'GraphAnnoFont' = ("<sans-serif>, <MTsans-serif>",10pt);
```

Its counterpart in the DEFAULT style is as follows:

```
class GraphFonts
   "Fonts used in graph styles" /
   'GraphDataFont' = ("<sans-serif>, <MTsans-serif>",7pt)
   'GraphUnicodeFont' = ("<MTsans-serif-unicode>",9pt)
   'GraphValueFont' = ("<sans-serif>, <MTsans-serif>",9pt)
   'GraphLabelFont' = ("<sans-serif>, <MTsans-serif>",10pt,bold)
   'GraphFootnoteFont' = ("<sans-serif>, <MTsans-serif>",10pt)
   'GraphTitleFont' = ("<sans-serif>, <MTsans-serif>",11pt,bold)
   'GraphAnnoFont' = ("<sans-serif>, <MTsans-serif>",10pt);
```

`GraphDataFont` is used for point and line labels, `GraphUnicodeFont` is used for special symbols, `GraphValueFont` is used for tick and legend labels, `GraphLabelFont` is used for axis labels, `GraphFootnoteFont` is used for footnotes, `GraphTitleFont` is used for titles, and `GraphAnnoFont` is used for annotations. Axis labels are bold in the DEFAULT style, whereas axis labels are not bold in the STATISTICAL style. Footnotes are italic in the DEFAULT style, whereas footnotes are not italic in the STATISTICAL style. Both styles use a bold font for the title, and that font is larger than the other fonts.

It is instructive to see the fonts that are used for tables and other nongraphical output in the RTF style as well, since that style element shows additional font specifications. The `fonts` style element in the RTF style is as follows:

```
style fonts /
   'TitleFont2' = ("<serif>, Times Roman",12pt,bold italic)
   'TitleFont' = ("<serif>, Times Roman",13pt,bold italic)
   'StrongFont' = ("<serif>, Times Roman",10pt,bold)
   'EmphasisFont' = ("<serif>, Times Roman",10pt,italic)
   'FixedEmphasisFont' = ("<monospace>, Courier",9pt,italic)
   'FixedStrongFont' = ("<monospace>, Courier",9pt,bold)
   'FixedHeadingFont' = ("<monospace>, Courier",9pt,bold)
   'BatchFixedFont' = ("SAS Monospace, <monospace>, Courier",6.7pt)
   'FixedFont' = ("<monospace>, Courier",9pt)
   'headingEmphasisFont' = ("<serif>, Times Roman",11pt,bold italic)
   'headingFont' = ("<serif>, Times Roman",11pt,bold)
   'docFont' = ("<serif>, Times Roman",10pt);
```

If you like the STATISTICAL style, but prefer the fonts from the DEFAULT style, you can make a new style called MYSTYLE as follows:

```
proc template;
   define style mystyle;
      parent = styles.statistical;
      class GraphFonts
         "Fonts used in graph styles" /
         'GraphDataFont' = ("<sans-serif>, <MTsans-serif>",7pt)
         'GraphUnicodeFont' = ("<MTsans-serif-unicode>",9pt)
         'GraphValueFont' = ("<sans-serif>, <MTsans-serif>",9pt)
         'GraphLabelFont' = ("<sans-serif>, <MTsans-serif>",10pt,bold)
         'GraphFootnoteFont' = ("<sans-serif>, <MTsans-serif>",10pt)
         'GraphTitleFont' = ("<sans-serif>, <MTsans-serif>",11pt,bold)
         'GraphAnnoFont' = ("<sans-serif>, <MTsans-serif>",10pt);
   end;
run;
```

You can use the following steps to compare the three styles:

```
ods listing style=default;

proc sgplot data=sashelp.class;
   title 'Class';
   footnote 'Default Style';
   reg y=weight x=height / datalabel=name;
run;

ods listing style=statistical;

proc sgplot data=sashelp.class;
   title 'Class';
   footnote 'Statistical Style';
   reg y=weight x=height / datalabel=name;
run;

ods listing style=mystyle;

proc sgplot data=sashelp.class;
   title 'Class';
   footnote 'MyStyle Style';
   reg y=weight x=height / datalabel=name;
run;

ods listing;
```

The results are displayed in Figure 3.1, Figure 3.2, and Figure 3.3.

Figure 3.1   Default Style

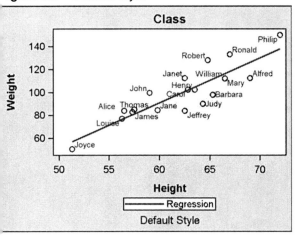

Figure 3.2   STATISTICAL Style

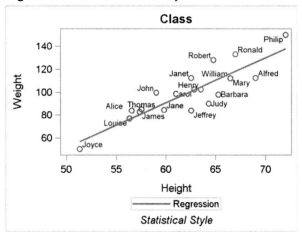

Figure 3.3   MyStyle Style

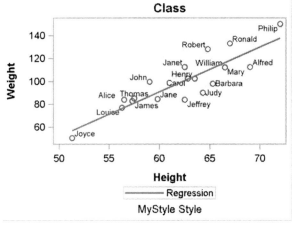

Figure 3.4   MyBold Style

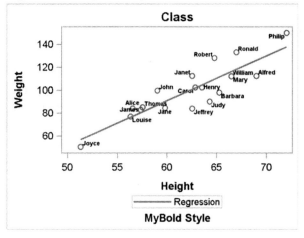

Alternatively, you could just modify some of the elements rather than redefining them all as follows:

```
proc template;
   define style mybold;
      parent = styles.statistical;
      style GraphFonts from GraphFonts /
         'GraphDataFont' = ("<sans-serif>, <MTsans-serif>",5pt,bold)
         'GraphLabelFont' = ("<sans-serif>, <MTsans-serif>",10pt,bold)
         'GraphFootnoteFont' = ("<sans-serif>, <MTsans-serif>",10pt,bold);
   end;
run;

ods listing style=mybold;

proc sgplot data=sashelp.class;
   title 'Class';
   footnote 'MyBold Style';
   reg y=weight x=height / datalabel=name;
run;

ods listing;
footnote;
```

The results are displayed in Figure 3.4. The statement STYLE *style-element* FROM *existing-style-element* creates a style element that inherits some of its features from an existing style element while overriding some of those features. It is equivalent to the statement: CLASS *style-element*. The MYBOLD style sets the font for data values (point labels) to bold and to a smaller-than-the-default, five-point font. The axis label and footnote fonts are also set to bold.

The following step shows how to change the type of font:

```
proc template;
   define style myfonts;
      parent = styles.statistical;
      style GraphFonts from GraphFonts /
         'GraphDataFont' = ("Arial",7pt)
         'GraphLabelFont' = ("Times New Roman",10pt)
         'GraphFootnoteFont' = ("Courier New",10pt);
   end;
run;
```

No results from using this style are displayed.

The next part of the STATISTICAL style that pertains to graphics is the following declaration of colors:

```
style GraphColors /
    'gablock' = cxF5F5F0
    'gblock' = cxDFE6EF
    'gcclipping' = cxDC531F
    'gclipping' = cxE7774F
    'gcstars' = cx445694
    'gstars' = cxCAD5E5
    'gcruntest' = cxE7774F
    'gruntest' = cxE6E6CC
    'gccontrollim' = cxCCCC97
    'gcontrollim' = cxFFFFE3
    'gdata' = cxCAD5E5
    'gcdata' = cx445694
    'goutlier' = cxB9CFE7
    'gcoutlier' = cx000000
    'gfit2' = cxDC531F
    'gfit' = cx667FA2
    'gcfit2' = cxDC531F
    'gcfit' = cx667FA2
    'gconfidence2' = cxE3D5CD
    'gconfidence' = cxB9CFE7
    'gcconfidence2' = cxE3D5CD
    'gcconfidence' = cxE3D5CD
    'gpredict' = cx667FA2
    'gcpredict' = cx445694
    'gpredictlim' = cx7486C4
    'gcpredictlim' = cx7486C4
    'gerror' = cxCA5E3D
    'gcerror' = cxA33708
    'greferencelines' = cxA5A5A5
    'gheader' = colors('docbg')
    'gconramp3cend' = cxFF3A2E
    'gconramp3cneutral' = cxEBC79E
    'gconramp3cstart' = cx445694
    'gramp3cend' = cx667FA2
    'gramp3cneutral' = cxFFFFFF
    'gramp3cstart' = cxAFB5A6
    'gconramp2cend' = cxA23A23
    'gconramp2cstart' = cxFFF1EF
    'gramp2cend' = cx445694
    'gramp2cstart' = cxF3F5FC
    'gtext' = cx000000
    'glabel' = cx000000
    'gborderlines' = cxD1D1D1
    'goutlines' = cx000000
    'ggrid' = cxE6E6E6
    'gaxis' = cxA5A5A5
    'gshadow' = cx8F8F8F
    'gfloor' = cxDCDAC9
    'glegend' = cxFFFFFF
    'gwalls' = cxFFFFFF
```

```
'gcdata12' = cxF9DA04
'gdata12' = cxDDD17E
'gcdata11' = cxB38EF3
'gdata11' = cxB7AEF1
'gcdata10' = cx47A82A
'gdata10' = cx87C873
'gcdata9' = cxD17800
'gdata9' = cxCF974B
'gcdata8' = cxB26084
'gdata8' = cxCD7BA1
'gcdata7' = cx2597FA
'gdata7' = cx94BDE1
'gcdata6' = cx7F8E1F
'gdata6' = cxBABC5C
'gcdata5' = cx9D3CDB
'gdata5' = cxB689CD
'gcdata4' = cx543005
'gdata4' = cxA9865B
'gcdata3' = cx01665E
'gdata3' = cx66A5A0
'gcdata2' = cxA23A2E
'gdata2' = cxD05B5B
'gcdata1' = cx445694
'gdata1' = cx6F7EB3;
```

There are two types of color elements. Many of the color elements begin with gc where g stands for graph and c stands for contrast. Contrast colors apply to markers and lines. Colors that begin with g for graph but that are without the c apply to filled areas such as confidence bands and contour plots.

These style elements control the colors of typical points in graphs (gdata, gcdata), outliers (goutlier, gcoutlier), regression fit functions or normal density functions (gfit, gcfit), secondary fit functions or kernel density functions (gfit2, gcfit2), confidence limits (gconfidence2, gconfidence, gcconfidence2, gcconfidence), prediction limits (gpredict, gcpredict, gpredictlim, gcpredictlim), reference lines (greferencelines), three-color ramps for contour plots and continuous legends (gconramp3cend, gconramp3cneutral, gconramp3cstart, gramp3cend, gramp3cneutral, gramp3cstart), two color ramps (gconramp2cend, gconramp2cstart, gramp2cend, gramp2cstart), and the first through twelfth group of observations with the GROUP= option (gcdata1, gdata1, gcdata2, gdata2, ..., gcdata11, gdata11, gcdata12, gdata12).

Figure 3.5 STATISTICAL Style

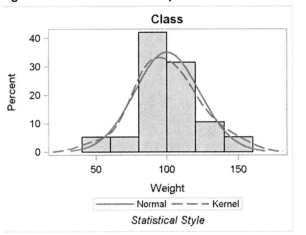

Figure 3.6 MyColor Style

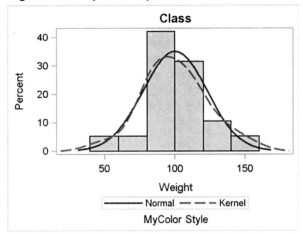

The following steps make two changes to the STATISTICAL style:

```
proc template;
   define style mycolor;
      parent = styles.statistical;
      class GraphFonts /
         'GraphFootnoteFont' = ("<sans-serif>, <MTsans-serif>",10pt);
      class GraphColors /
         'gcfit'  = blue
         'gcfit2' = red;
   end;
run;
```

The default font for footnotes is replaced by a nonitalic font, and the contrast colors that are used for the two fit lines are replaced by a pure blue and a pure red. The following steps show two density estimates using both the STATISTICAL style and the new MYCOLOR style:

```
ods listing style=statistical;

proc sgplot data=sashelp.class;
   title 'Class';
   footnote 'Statistical Style';
   histogram weight;
   density weight / type=normal;
   density weight / type=kernel;
run;
```

```
ods listing style=mycolor;

proc sgplot data=sashelp.class;
   title 'Class';
   footnote 'MyColor Style';
   histogram weight;
   density weight / type=normal;
   density weight / type=kernel;
run;

ods listing;
footnote;
```

The results are displayed in Figure 3.5 and Figure 3.6.

PROC SGPLOT generates a template using the GTL for producing this graph. Two of the statements that it generates are as follows:

```
DensityPlot Weight / Lineattrs=GraphFit normal()
                     LegendLabel="Normal" NAME="DENSITY";
DensityPlot Weight / Lineattrs=GraphFit2 kernel()
                     LegendLabel="Kernel" NAME="DENSITY1";
```

By default, PROC SGPLOT recognizes that the two DENSITY statements produce functions that need to be distinguished, and it uses the `GraphFit` and `GraphFit2` style elements to set the line attributes.

There are no other graphical style elements defined in the STATISTICAL style. In particular, the style elements `GraphFit` and `GraphFit2` are not defined. Their definitions and many others are in the parent DEFAULT style. It contains the following two statements:

```
class GraphFit /
   linethickness = 2px
   linestyle = 1
   markersize = 7px
   markersymbol = "circle"
   contrastcolor = GraphColors('gcfit')
   color = GraphColors('gfit');

class GraphFit2 /
   linethickness = 2px
   linestyle = 4
   markersize = 7px
   markersymbol = "X"
   contrastcolor = GraphColors('gcfit2')
   color = GraphColors('gfit2');
```

The first DENSITYPLOT statement specifies the option LINEATTRS= and references the `GraphFit` style element. The DEFAULT style specifies the definition of the `GraphFit` style element, and it is not overridden by either the STATISTICAL style or the MYCOLOR style. The DEFAULT style sets the contrast color (the line color) for the fit line as follows: `contrastcolor = GraphColors('gcfit')`. The value of `gcfit` defined in `GraphColors` is used. The DEFAULT style sets this color as follows: `'gcfit'` = `cx003178`. The STATISTICAL style overrides the DEFAULT style definition and sets the color as follows: `'gcfit'` = `cx667FA2`. The MYCOLOR style overrides the STATISTICAL style definition and sets the color as follows: `'gcfit'` = `blue`. That is how the line becomes blue. The explanation for how the second fit line becomes red is analogous.

You could have gotten the same results with the following steps:

```
proc template;
   define style mycolor;
      parent = styles.statistical;
      class GraphFonts /
         'GraphFootnoteFont' = ("<sans-serif>, <MTsans-serif>",10pt);
      class GraphFit  / contrastcolor = blue;
      class GraphFit2 / contrastcolor = red;
   end;
run;

ods listing style=mycolor;
proc sgplot data=sashelp.class;
   title 'Class';
   footnote 'MyColor Style';
   histogram weight;
   density weight / type=normal;
   density weight / type=kernel;
run;

footnote;
```

The results of these steps are not shown, but they exactly match Figure 3.6. In this case, the two versions of the MYCOLOR style are exactly equivalent. However, if the `gcfit` and `gcfit2` colors had been used in more than one place, the two versions would not have been equivalent.

Part of the challenge of style modification is knowing what to change. The following step produces two linear regression fit lines, one for males and one for females:

```
ods listing style=mycolor;
proc sgplot data=sashelp.class;
   reg y=weight x=height / datalabel=name group=sex;
run;
ods listing;
```

The results of this step are not shown, but they are not affected by this style change. Since the two fit functions come from a GROUP= option, it is the `GraphData1` and `GraphData2` style elements that control the appearance of the two groups, not the `GraphFit` and `GraphFit2` style elements.

In this section, we have seen that the STATISTICAL style sets colors, but it does not change the style elements that use those colors. Rather, it relies on those style elements picking up the colors indirectly from the `GraphColor` style element. Similarly, the `GraphFonts` style element is defined in the STATISTICAL style, and it is used by the DEFAULT style in style elements such as the following:

```
class GraphTitleText /
   font = GraphFonts('GraphTitleFont')
   color = GraphColors('gtext');
```

The only way to find out for sure how a style is defined is to use PROC TEMPLATE and the SOURCE statement to see it and its parent(s). However, the following steps display the major graphical components of the DEFAULT style and hence the STATISTICAL style. In some cases, comments are added about ways in which you might want to modify the style. In most cases, you should want to leave the style alone. However, if you really want to change something, these next sections will help. Subsequent steps provide information on a few of the other styles that are commonly used for statistical work.

3.2 ODS Style Modification

This section provides a series of steps for modifying discrete parts of the STATISTICAL style. You can change the PARENT= option to modify other styles.

You can modify the graph wall or background area bounded by the axes by modifying the following style definition:

```
proc template;
   define style mystyle;
      parent = styles.statistical;
      class GraphWalls /
         linethickness = 1px
         linestyle = 1
         frameborder = on
         contrastcolor = GraphColors('gaxis')
         backgroundcolor = GraphColors('gwalls')
         color = GraphColors('gwalls');
   end;
run;
```

You can modify the axes by modifying the following style definition:

```
proc template;
   define style mystyle;
      parent = styles.statistical;
      class GraphAxisLines /
         tickdisplay = "outside"
         linethickness = 1px
         linestyle = 1
         contrastcolor = GraphColors('gaxis')
         color = GraphColors('gaxis');
   end;
run;
```

You can modify the grid lines by modifying the following style definition:

```
proc template;
   define style mystyle;
      parent = styles.statistical;
      class GraphGridLines /
         displayopts = "auto"
         linethickness = 1px
         linestyle = 1
         contrastcolor = GraphColors('ggrid')
         color = GraphColors('ggrid');
   end;
run;
```

The LAYOUT OVERLAY statement has the options: `xAxisOpts=(GridDisplay=Auto_On)` `yAxisOpts=(GridDisplay=Auto_On)`. Other values for GRIDDISPLAY= include ON and OFF. If you really like grids, you can modify the style template by specifying `displayopts = "on"` to display grids everywhere except when `GridDisplay=Off` is specified in the graph template. If you are not fond of grids, then you can modify the style template by specifying `displayopts = "off"` to suppress grids everywhere except when `GridDisplay=On` is specified in the graph template. The line style for grids is style 1, which is a solid line. You could specify `LineStyle = 2` for a dashed grid line. Many other line patterns are available.

You can modify outline properties for fill areas such as bars, box plots, ellipses, and histograms by modifying the following style definition:

```
proc template;
   define style mystyle;
      parent = styles.statistical;
      class GraphOutlines /
         linethickness = 1px
         linestyle = 1
         contrastcolor = GraphColors('goutlines')
         color = GraphColors('goutlines');
   end;
run;
```

You can modify border lines, such as the border around the graph wall and the legend border by modifying the following style definition:

```
proc template;
   define style mystyle;
      parent = styles.statistical;
      class GraphBorderLines /
         linethickness = 1px
         linestyle = 1
         contrastcolor = GraphColors('gborderlines')
         color = GraphColors('gborderlines');
   end;
run;
```

You can eliminate the borders by setting `LineThickness = 0px`.

You can modify reference lines by modifying the following style definition:

```
proc template;
   define style mystyle;
      parent = styles.statistical;
      class GraphReference /
         linethickness = 1px
         linestyle = 1
         contrastcolor = GraphColors('greferencelines');
   end;
run;
```

You can modify the appearance of the title by modifying the following style definition:

```
proc template;
   define style mystyle;
      parent = styles.statistical;
      class GraphTitleText /
         font = GraphFonts('GraphTitleFont')
         color = GraphColors('gtext');
   end;
run;
```

You can modify the appearance of the footnote by modifying the following style definition:

```
proc template;
   define style mystyle;
      parent = styles.statistical;
      class GraphFootnoteText /
         font = GraphFonts('GraphFootnoteFont')
         color = GraphColors('gtext');
   end;
run;
```

You can modify the point and line labels by modifying the following style definition:

```
proc template;
   define style mystyle;
      parent = styles.statistical;
      class GraphDataText /
         font = GraphFonts('GraphDataFont')
         color = GraphColors('gtext');
   end;
run;
```

You can modify the axis labels and legend titles by modifying the following style definition:

```
proc template;
   define style mystyle;
      parent = styles.statistical;
      class GraphLabelText /
         font = GraphFonts('GraphLabelFont')
         color = GraphColors('glabel');
   end;
run;
```

You can modify the font for tick and legend values by modifying the following style definition:

```
proc template;
   define style mystyle;
      parent = styles.statistical;
      class GraphValueText /
         font = GraphFonts('GraphValueFont')
         color = GraphColors('gtext');
   end;
run;
```

You can modify the Unicode text for special characters by modifying the following style definition:

```
proc template;
   define style mystyle;
      parent = styles.statistical;
      class GraphUnicodeText /
         font = GraphFonts('GraphUnicodeFont');
   end;
run;
```

You can modify the graph background by modifying the following style definition:

```
proc template;
   define style mystyle;
      parent = styles.statistical;
      class GraphBackground /
         backgroundcolor = colors('docbg')
         color = colors('docbg');
   end;
run;
```

You can modify the floor of 3D plots by modifying the following style definition:

```
proc template;
   define style mystyle;
      parent = styles.statistical;
      class GraphFloor /
         backgroundcolor = GraphColors('gfloor')
         color = GraphColors('gfloor');
   end;
run;
```

You can modify the legend background by modifying the following style definition:

```
proc template;
   define style mystyle;
      parent = styles.statistical;
      class GraphLegendBackground /
         backgroundcolor = GraphColors('glegend')
         color = GraphColors('glegend');
   end;
run;
```

You can modify the background for the legend title by modifying the following style definition:

```
proc template;
   define style mystyle;
      parent = styles.statistical;
      class GraphHeaderBackground /
         backgroundcolor = GraphColors('gheader')
         color = GraphColors('gheader');
   end;
run;
```

You can modify the lines, markers, and colors for nongrouped data items by modifying the following style definition:

```
proc template;
   define style mystyle;
      parent = styles.statistical;
      class GraphDataDefault /
         endcolor = GraphColors('gramp3cend')
         neutralcolor = GraphColors('gramp3cneutral')
         startcolor = GraphColors('gramp3cstart')
         markersize = 7px
         markersymbol = "circle"
         linethickness = 1px
         linestyle = 1
         contrastcolor = GraphColors('gcdata')
         color = GraphColors('gdata');
   end;
run;
```

When you have a GROUP= option, you can modify the displays of group 1 through group 12 by modifying the `GraphData1` through `GraphData12` style definitions. In the following twelve definitions, there are seven marker symbols defined, 11 line styles, and 12 colors. These style elements cycle as the number groups increases. For example, the eighth group uses the `GraphData1` marker symbol. If you have more than 12 groups, you can define additional `GraphData`n style elements. This is discussed later in this section.

You can modify the display of group 1 by modifying the following style definition:

```
proc template;
   define style mystyle;
      parent = styles.statistical;
      class GraphData1 /
         markersymbol = "circle"
         linestyle = 1
         contrastcolor = GraphColors('gcdata1')
         color = GraphColors('gdata1');
   end;
run;
```

See Figure 3.7 and Figure 3.8 for an example that displays the `GraphData`n and several other style elements.

You can modify the display of group 2 by modifying the following style definition:

```
proc template;
   define style mystyle;
      parent = styles.statistical;
      class GraphData2 /
         markersymbol = "plus"
         linestyle = 4
         contrastcolor = GraphColors('gcdata2')
         color = GraphColors('gdata2');
      end;
   run;
```

You can modify the display of group 3 by modifying the following style definition:

```
proc template;
   define style mystyle;
      parent = styles.statistical;
      class GraphData3 /
         markersymbol = "X"
         linestyle = 8
         contrastcolor = GraphColors('gcdata3')
         color = GraphColors('gdata3');
      end;
   run;
```

You can modify the display of group 4 by modifying the following style definition:

```
proc template;
   define style mystyle;
      parent = styles.statistical;
      class GraphData4 /
         markersymbol = "triangle"
         linestyle = 5
         contrastcolor = GraphColors('gcdata4')
         color = GraphColors('gdata4');
      end;
   run;
```

You can modify the display of group 5 by modifying the following style definition:

```
proc template;
   define style mystyle;
      parent = styles.statistical;
      class GraphData5 /
         markersymbol = "square"
         linestyle = 14
         contrastcolor = GraphColors('gcdata5')
         color = GraphColors('gdata5');
      end;
   run;
```

You can modify the display of group 6 by modifying the following style definition:

```
proc template;
   define style mystyle;
      parent = styles.statistical;
      class GraphData6 /
         markersymbol = "asterisk"
         linestyle = 26
         contrastcolor = GraphColors('gcdata6')
         color = GraphColors('gdata6');
      end;
   run;
```

You can modify the display of group 7 by modifying the following style definition:

```
proc template;
   define style mystyle;
      parent = styles.statistical;
      class GraphData7 /
         markersymbol = "diamond"
         linestyle = 15
         contrastcolor = GraphColors('gcdata7')
         color = GraphColors('gdata7');
      end;
   run;
```

You can modify the display of group 8 by modifying the following style definition:

```
proc template;
   define style mystyle;
      parent = styles.statistical;
      class GraphData8 /
         linestyle = 20
         contrastcolor = GraphColors('gcdata8')
         color = GraphColors('gdata8');
      end;
   run;
```

You can modify the display of group 9 by modifying the following style definition:

```
proc template;
   define style mystyle;
      parent = styles.statistical;
      class GraphData9 /
         linestyle = 41
         contrastcolor = GraphColors('gcdata9')
         color = GraphColors('gdata9');
      end;
   run;
```

You can modify the display of group 10 by modifying the following style definition:

```
proc template;
   define style mystyle;
      parent = styles.statistical;
      class GraphData10 /
         linestyle = 42
         contrastcolor = GraphColors('gcdata10')
         color = GraphColors('gdata10');
   end;
run;
```

You can modify the display of group 11 by modifying the following style definition:

```
proc template;
   define style mystyle;
      parent = styles.statistical;
      class GraphData11 /
         linestyle = 2
         contrastcolor = GraphColors('gcdata11')
         color = GraphColors('gdata11');
   end;
run;
```

You can modify the display of group 12 by modifying the following style definition:

```
proc template;
   define style mystyle;
      parent = styles.statistical;
      class GraphData12 /
         contrastcolor = GraphColors('gcdata12')
         color = GraphColors('gdata12');
   end;
run;
```

You can modify the two-color ramp, which is used with gradient contours, surfaces, markers, and data labels with a continuous color response, by modifying the following style definition:

```
proc template;
   define style mystyle;
      parent = styles.statistical;
      class TwoColorRamp /
         endcolor = GraphColors('gramp2cend')
         startcolor = GraphColors('gramp2cstart');
   end;
run;
```

You can modify the three-color ramp, which is used with gradient contours, surfaces, markers, and data labels with a continuous color response, by modifying the following style definition:

```
proc template;
   define style mystyle;
      parent = styles.statistical;
      class ThreeColorRamp /
         endcolor = GraphColors('gramp3cend')
         neutralcolor = GraphColors('gramp3cneutral')
         startcolor = GraphColors('gramp3cstart');
      end;
   run;
```

You can modify the alternative three-color ramp, which is used with line contours, markers, and data labels with a segmented range color response, by modifying the following style definition:

```
proc template;
   define style mystyle;
      parent = styles.statistical;
      class ThreeColorAltRamp /
         endcolor = GraphColors('gconramp3cend')
         neutralcolor = GraphColors('gconramp3cneutral')
         startcolor = GraphColors('gconramp3cstart');
      end;
   run;
```

You can modify the properties of outliers by modifying the following style definition:

```
proc template;
   define style mystyle;
      parent = styles.statistical;
      class GraphOutlier /
         linethickness = 2px
         linestyle = 42
         markersize = 7px
         markersymbol = "circle"
         contrastcolor = GraphColors('gcoutlier')
         color = GraphColors('goutlier');
      end;
   run;
```

You can modify the primary fit function by modifying the following style definition:

```
proc template;
   define style mystyle;
      parent = styles.statistical;
      class GraphFit /
         linethickness = 2px
         linestyle = 1
         markersize = 7px
         markersymbol = "circle"
         contrastcolor = GraphColors('gcfit')
         color = GraphColors('gfit');
      end;
   run;
```

You can modify the secondary fit function by modifying the following style definition:

```
proc template;
   define style mystyle;
      parent = styles.statistical;
      class GraphFit2 /
         linethickness = 2px
         linestyle = 4
         markersize = 7px
         markersymbol = "X"
         contrastcolor = GraphColors('gcfit2')
         color = GraphColors('gfit2');
      end;
   run;
```

You can modify the primary fit confidence interval by modifying the following style definition:

```
proc template;
   define style mystyle;
      parent = styles.statistical;
      class GraphConfidence /
         linethickness = 1px
         linestyle = 1
         markersize = 7px
         markersymbol = "triangle"
         contrastcolor = GraphColors('gcconfidence')
         color = GraphColors('gconfidence');
      end;
   run;
```

You can modify the secondary fit confidence interval by modifying the following style definition:

```
proc template;
   define style mystyle;
      parent = styles.statistical;
      class GraphConfidence2 /
         linethickness = 1px
         linestyle = 4
         markersize = 7px
         markersymbol = "diamond"
         contrastcolor = GraphColors('gcconfidence2')
         color = GraphColors('gconfidence2');
      end;
   run;
```

You can modify prediction lines by modifying the following style definition:

```
proc template;
   define style mystyle;
      parent = styles.statistical;
      class GraphPrediction /
         linethickness = 2px
         linestyle = 4
         markersize = 7px
         markersymbol = "plus"
         contrastcolor = GraphColors('gcpredict')
         color = GraphColors('gpredict');
      end;
   run;
```

You can modify the prediction limits by modifying the following style definition:

```
proc template;
   define style mystyle;
      parent = styles.statistical;
      class GraphPredictionLimits /
         linethickness = 1px
         linestyle = 2
         markersize = 7px
         markersymbol = "chain"
         contrastcolor = GraphColors('gcpredictlim')
         color = GraphColors('gpredictlim');
      end;
   run;
```

You can modify the error line or error bar fill by modifying the following style definition:

```
proc template;
   define style mystyle;
      parent = styles.statistical;
      class GraphError /
         linethickness = 1px
         linestyle = 1
         markersize = 7px
         markersymbol = "asterisk"
         contrastcolor = GraphColors('gcerror')
         color = GraphColors('gerror');
   end;
run;
```

You can modify the display options for box plots by modifying the following style definition:

```
proc template;
   define style mystyle;
      parent = styles.statistical;
      class GraphBox /
         capstyle = "serif"
         connect = "mean"
         displayopts = "fill caps median mean outliers";
   end;
run;
```

You can modify the box plot median display by modifying the following style definition:

```
proc template;
   define style mystyle;
      parent = styles.statistical;
      class GraphBoxMedian /
         linestyle = 1
         linethickness = 1px
         contrastcolor = GraphColors('gcdata');
   end;
run;
```

You can modify the box plot mean display by modifying the following style definition:

```
proc template;
   define style mystyle;
      parent = styles.statistical;
      class GraphBoxMean /
         markersize = 9px
         markersymbol = "diamond"
         contrastcolor = GraphColors('gcdata');
   end;
run;
```

You can modify the box plot whisker display by modifying the following style definition:

```
proc template;
   define style mystyle;
      parent = styles.statistical;
      class GraphBoxWhisker /
         linestyle = 1
         linethickness = 1px
         contrastcolor = GraphColors('gcdata');
      end;
   run;
```

You can modify the histogram composition by modifying the following style definition:

```
proc template;
   define style mystyle;
      parent = styles.statistical;
      class GraphHistogram /
         displayopts = "fill outline";
      end;
   run;
```

You can modify ellipses by modifying the following style definition:

```
proc template;
   define style mystyle;
      class GraphEllipse /
         displayopts = "outline";
      parent = styles.statistical;
      end;
   run;
```

You can modify band plots by modifying the following style definition:

```
proc template;
   define style mystyle;
      parent = styles.statistical;
      class GraphBand /
         displayopts = "fill";
      end;
   run;
```

You can modify contour plots by modifying the following style definition:

```
proc template;
   define style mystyle;
      parent = styles.statistical;
      class GraphContour /
         displayopts = "LabeledLineGradient";
      end;
   run;
```

You can modify block plots by modifying the following style definition:

```
proc template;
   define style mystyle;
      parent = styles.statistical;
      class GraphBlock /
         color = GraphColors('gblock');
   end;
run;
```

You can modify the alternative block plot display by modifying the following style definition:

```
proc template;
   define style mystyle;
      parent = styles.statistical;
      class GraphAltBlock /
         color = GraphColors('gablock');
   end;
run;
```

You can modify annotation lines by modifying the following style definition:

```
proc template;
   define style mystyle;
      parent = styles.statistical;
      class GraphAnnoLine /
         linestyle = 1
         linethickness = 2px
         contrastcolor = GraphColors('gcdata');
   end;
run;
```

You can modify annotation text by modifying the following style definition:

```
proc template;
   define style mystyle;
      parent = styles.statistical;
      class GraphAnnoText /
         font = GraphFonts('GraphAnnoFont')
         color = GraphColors('gtext');
   end;
run;
```

You can modify annotation shapes by modifying the following style specification:

```
proc template;
   define style mystyle;
      class GraphAnnoShape /
         markersize = 12px
         markersymbol = "starfilled"
         linethickness = 2px
         linestyle = 1
         contrastcolor = GraphColors('gcdata')
         color = GraphColors('gdata');
      parent = styles.statistical;
   end;
run;
```

The DEFAULT style contains other graphical style elements and many nongraphical style elements as well. You can use PROC TEMPLATE with the SOURCE statement to see them all. In most ways, the STATISTICAL, LISTING, and ANALYSIS styles are similar to the DEFAULT STYLE, but they use different colors and fonts. The RTF style uses different fonts and a few different colors.

3.3   Colors and Groups

This section discusses and displays colors, the **GraphData**n style elements, and how they are used to display colors in graphs of data that are grouped.

The JOURNAL style is a noncolor style designed for use in black and white media such as journals and noncolor printers. Hence its group colors are all shades of gray. When the red, green, and blue components are all set to 00, the color is black, when they are all set to FF, the color is white, and when they are otherwise all the same, the color is a shade of gray. The JOURNAL style group colors are: 'gdata1' = CXbfbfbf, 'gdata2' = CX828282, 'gdata3' = CX595959, 'gdata4' = CX303030, 'gdata5' = CXcfcfcf, 'gdata6' = CX1c1c1c, 'gdata7' = CXababab, 'gdata8' = CX969696, 'gdata9' = CX6e6e6e, 'gdata10' = CX454545, 'gdata11' = CXe1e1e1, 'gdata12' = CX080808, and its contrast colors are all black (cx000000). The JOURNAL style also redefines all of the **GraphData**n style elements so that each one has a different symbol. Recall from the DEFAULT style that markers are not defined for all of the **GraphData**n style elements. A portion of the JOURNAL style is as follows:

```
style GraphData1 from GraphData1 /
   markersymbol = "circle";
style GraphData2 from GraphData2 /
   markersymbol = "plus";
   .
   .
   .
style GraphData12 from GraphData12 /
   markersymbol = "tack";
```

The entire definition of the JOURNAL2 style is as follows:

```
define style Styles.Journal2;
   parent = styles.journal;
   style GraphHistogram from GraphComponent /
      displayopts = "outline";
   style GraphEllipse from GraphComponent /
      displayopts = "outline";
   style GraphBand from GraphComponent /
      displayopts = "outline";
   style GraphBox from GraphComponent /
      displayopts = "caps median mean outliers"
      connect = "mean"
      capstyle = "serif";
end;
```

It inherits from the JOURNAL style, which inherits from the DEFAULT style. It differs from the JOURNAL style in that the JOURNAL2 style only provides outlines for areas that are filled in the JOURNAL style.

You can use the following program to see the color and other attributes for a number of style elements in the DEFAULT style:

```
proc format;
   value vf 5 = 'GraphValueText';
run;

data x;
   array y[20] y0 - y19;
   do x = 1 to 20; y[x] = x - 0.5; end;
   do x = 0 to 10 by 5; output; end;
   label y0 = 'GraphLabelText' x = 'GraphLabelText';
   format x y0 vf.;
run;
```

Figure 3.7 Attributes of Style Elements in the DEFAULT Style

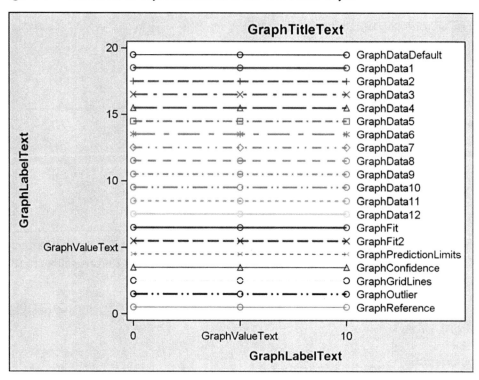

```
%macro l(i, 1);
   reg y=y&i x=x / lineattrs=&1 markerattrs=&1 curvelabel="  &1"
                   curvelabelpos=max;
%mend;

ods listing style=default;
proc sgplot noautolegend;
   title 'GraphTitleText';
   %macro d; %do i = 1 %to 12;
      reg y=y%eval(19-&i) x=x / lineattrs=GraphData&i markerattrs=GraphData&i
                                curvelabel="  GraphData&i" curvelabelpos=max;
   %end; %mend; %d
   %l(19, GraphDataDefault)
   %l( 6, GraphFit)
   %l( 5, GraphFit2)
   %l( 4, GraphPredictionLimits)
   %l( 3, GraphConfidence)
   %l( 2, GraphGridLines)
   %l( 1, GraphOutlier)
   %l( 0, GraphReference)
   xaxis values=(0 5 10);
run;
```

The results in Figure 3.7 display the attributes for a number of the elements of the DEFAULT style. Note that the marker symbols are defined for **GraphData1** through **GraphData7** style elements. Style elements **GraphData8** through **GraphData12** use the default circle from the **GraphDataDefault** style element. However, if a GROUP= variable was specified, then the marker symbols cycle, and group 8 displays a circle, group 9 displays a plus, and so on. This is explained further in the next example.

Figure 3.8 Markers and Lines Cycle with Different Periods in Groups

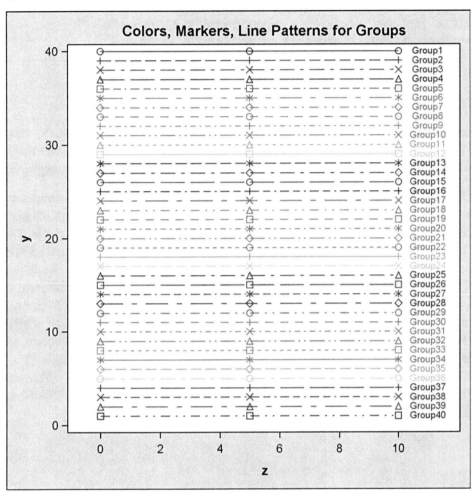

When there is a group or classification variable, the colors, markers, and lines that distinguish the groups are derived from the **GraphData***n* elements that are defined with the style. In the DEFAULT style, these are elements **GraphData1** through **GraphData12**. There can be any number of groups even though only twelve **GraphData***n* style elements are defined in the DEFAULT style. The following steps create a data set with 40 groups, display one line per group, and produce Figure 3.8:

```
data x;
   do y = 40 to 1 by -1;
      group = 'Group' || put(41 - y, 2. -L);
      do x = 0 to 10 by 5;
         if x = 10 then do; z = 11; l = group; end;
         else            do; z = .;  l = ' ';   end;
         output;
      end;
   end;
run;
```

```
proc sgplot data=x;
   title 'Colors, Markers, Line Patterns for Groups';
   series  y=y x=x / group=group markers;
   scatter y=y x=z / group=group markerchar=1;
run;

ods listing;
```

The colors, markers, and line patterns in Figure 3.8 repeat in cycles. The `GraphData1 − GraphData8` lines in Figure 3.7 exactly match the `Group1 − Group8` lines in Figure 3.8. After that, there are differences due to the cyclic construction of the grouped style definition. This is explained next.

The DEFAULT style defines a marker symbol only in `GraphData1` through `GraphData7`. The seven markers are: circle, plus, X, triangle, square, asterisk, and diamond. With the explicit style reference in Figure 3.7, the actual symbol (when no symbol is specified) is the circle. This is what you see for `GraphData8` through `GraphData12`. With the group variable in Figure 3.8, the symbols repeat in cycles. Hence, `Group1`, `Group8`, `Group15`, and so on are all circles. Similarly, `Group2`, `Group9`, `Group16`, and so on are all pluses. The DEFAULT style defines 11 different line styles for `GraphData1` through `GraphData11`: 1, 4, 8, 5, 14, 26, 15, 20, 41, 42, and 2. Hence, `Group1`, `Group12`, `Group23`, and so on all have the same line style, which is a solid line. Similarly, `Group2`, `Group13`, `Group24`, and so on all have line style 4. There are twelve different colors, so `Group1`, `Group13`, `Groupa25`, and so on all have the same colors. Overall, there are $12 \times 11 \times 7 = 924$ color/line/marker combinations that appear before any combination repeats. You can use the `ModStyle` SAS autocall macro (see the section "An All-Color Style" on page 145) to conveniently change these style attributes.

Figure 3.9 STATISTICAL Style

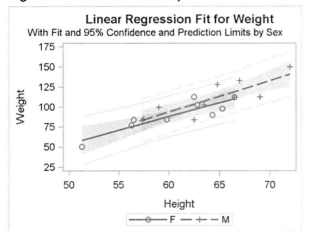

Figure 3.10 A Color-Based Style

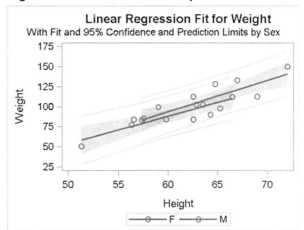

3.4 An All-Color Style

One particular set of style elements allows you to modify how groups of observations are distinguished in a graph. These are the **GraphData***n* style elements (**GraphData1** through **GraphData12**). In most cases, it is easiest to modify these elements through the **ModStyle** SAS autocall macro instead of directly modifying a style.[1] The first examples illustrate using this macro.

Many styles are designed to make color plots in which you can distinguish lines, functions, and groups of observations even when you send the plot to a black-and-white printer. Hence, lines and markers differ not only in color but also in pattern and symbol. You can use the **ModStyle** autocall macro to create a new style (for example, STATCOLOR) by modifying a parent style and reordering the colors, line patterns, and marker symbols in the **GraphData***n* style elements.

When you only specify a parent style and a new style name, and you use the defaults for all other options, the **ModStyle** macro creates a new style that uses only color to distinguish the groups. Lines and markers in subsequent groups match the lines and markers in the first group. The following example creates a plot with two groups, first with the STATISTICAL style, and then with a color-only style created by the default use of the **ModStyle** macro:

```
proc transreg data=sashelp.class;
   model identity(weight) = class(sex / zero=none) | identity(height);
run;

%modstyle(parent=statistical, name=StatColor)

ods listing style=StatColor;

proc transreg data=sashelp.class;
   model identity(weight) = class(sex / zero=none) | identity(height);
run;
```

---

[1]The **ModStyle** macro was written by Bob Derr at SAS, and information about it can be found in Kuhfeld (2009).

Figure 3.11 Gender Style

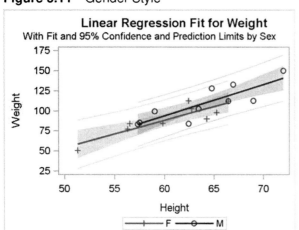

Figure 3.12 An Ad Hoc Style

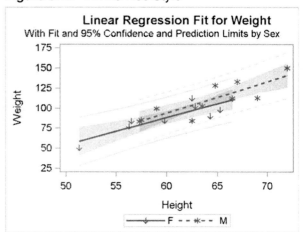

The graph using the STATISTICAL style is displayed in Figure 3.9, and the graph using the new style is displayed in Figure 3.10. In both plots, the females are represented by blue circles and a blue solid line. In the first plot, the males are represented by red pluses and a red dashed line, whereas in the second plot, the males and female groups differ only by color.

3.5   An Ad Hoc Style Modification for Groups

Styles are general; they are not made for specific types of data that have common color associations such as blue with male. The following step modifies the line styles, markers, and colors of the STATISTICAL style and creates a new style where the points for males are displayed in blue:

```
%modstyle(parent=statistical, name=GenderStyle, type=CLM,
          colors=red blue, fillcolors=red blue,
          markers=plus circle, linestyles=solid solid)
```

The colors for the first group (females, since F comes before M) are set to red, and the colors for the second group are set to blue.[2] The COLORS= option controls the line and marker colors, and the FILLCOLORS= option controls the colors for the confidence limits. The actual colors for the prediction and confidence limits are lighter due to the application of transparency. The marker for females is a plus, and the marker for males is a circle. The line style for both groups is a solid line. The TYPE=CLM option specifies that colors C, lines L, and markers M all vary simultaneously. The following step uses the new style and creates Figure 3.11:

```
ods listing style=GenderStyle;

proc transreg data=sashelp.class;
   model identity(weight) = class(sex / zero=none) | identity(height);
run;
```

---

[2]This example uses PROC TRANSREG, which sorts the levels of the CLASS variable into alphabetical order (F then M). PROC SGPLOT users will need to sort the data set to ensure that F comes before M.

The following step modifies the line styles, markers, and colors of the STATISTICAL style, creates a new style, and uses it to create Figure 3.12:

```
%modstyle(parent=statistical, name=GenderStyle, type=CLM,
         colors=GraphColors("gcdata1") GraphColors("gcdata2")
               green cx543005 cx9D3CDB cx7F8E1F cx2597FA cxB26084
               cxD17800 cx47A82A cxB38EF3 cxF9DA04 magenta,
         fillcolors=colors,
         linestyles=Solid ShortDash MediumDash LongDash MediumDashShortDash
               DashDashDot DashDotDot Dash LongDashShortDash Dot
               ThinDot ShortDashDot MediumDashDotDot,
         markers=ArrowDown Asterisk Circle CircleFilled Diamond
               DiamondFilled GreaterThan Hash HomeDown Ibeam Plus
               Square SquareFilled Star StarFilled Tack Tilde
               Triangle TriangleFilled Union X Y Z)

ods listing style=GenderStyle;

proc transreg data=sashelp.class;
   model identity(weight) = class(sex / zero=none) | identity(height);
run;
```

More style elements are changed than are displayed by the graph. However, the **ModStyle** macro options in this example illustrate some of the flexibility that you have for changing the **GraphData***n* elements. You can specify colors by name or by **cx***rrggbb* value, or you can use the colors that are predefined with the style. You can specify FILLCOLORS=COLORS when you want the fill colors to match what you specified for the COLORS= option. A number of different line styles and markers are available for you to use.

Figure 3.13 Survival Plot

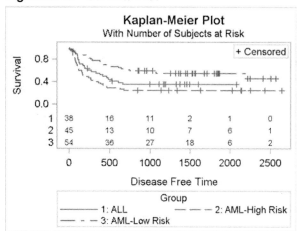

Figure 3.14 Modified Survival Plot

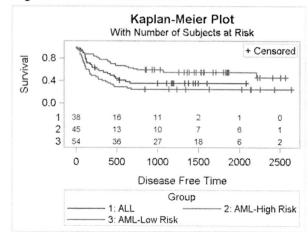

3.6   Color Changes in a Survival Plot

The section "Conditional Template Logic" on page 175 discussed changing the title of the survival estimate plot to "Kaplan-Meier Plot" by changing three ENTRYTITLE statements. This section assumes that change has been made. The following example uses PROC LIFETEST to produce a survival plot with the number of subjects at risk and multiple comparisons of survival curves:

```
proc format;
    value risk 1='ALL' 2='AML-Low Risk' 3='AML-High Risk';
run;

data BMT;
    input Group T Status @@;
    format Group risk.;
    label T='Disease Free Time';
    datalines;
1 2081 0 1 1602 0 1 1496 0 1 1462 0 1 1433 0
1 1377 0 1 1330 0 1  996 0 1  226 0 1 1199 0

    ... more lines ...

3  113 1 3  363 1
;

proc lifetest data=BMT plots=survival(atrisk=0 to 2500 by 500);
    time T * Status(0);
    strata Group / test=logrank adjust=sidak;
run;
```

The results are displayed in Figure 3.13.

The following steps use the **ModStyle** macro to change the colors of the survival curves to pure shades of blue, red, and green and to re-create the plot:

```
%modstyle(parent=statistical, name=SurvivalStyle,
          colors=blue red green)

ods listing style=SurvivalStyle;

proc lifetest data=BMT plots=survival(atrisk=0 to 2500 by 500);
   time T * Status(0);
   strata Group / test=logrank adjust=sidak;
run;
```

The results are displayed in Figure 3.14.

3.7 Direct Style Modifications

This example directly modifies a style template. You begin by using PROC TEMPLATE to display the style that you want to change:

```
proc template;
   source styles.statistical;
run;
```

The first two lines of the results are as follows:

```
define style Styles.Statistical;
   parent = styles.default;
```

The STATISTICAL style inherits many of its elements from the DEFAULT style. The preceding lines are followed by many lines that specify differences between the parent DEFAULT style and the STATISTICAL style. To see more fully how the STATISTICAL style is defined, you also need to look at its parent, as follows:

```
proc template;
   source styles.default;
run;
```

A small part of the results are as follows:

```
class GraphColors
   'gcdata' = cx000000
   'gdata' = cxB9CFE7
   . . .
   'gcdata1' = cx2A25D9
   . . .
   'gdata1' = cx7C95CA;
   . . .
```

```
class GraphDataDefault /
   endcolor = GraphColors('gramp3cend')
   neutralcolor = GraphColors('gramp3cneutral')
   startcolor = GraphColors('gramp3cstart')
   markersize = 7px
   markersymbol = "circle"
   linethickness = 1px
   linestyle = 1
   contrastcolor = GraphColors('gcdata')
   color = GraphColors('gdata');
class GraphData1 /
   markersymbol = "circle"
   linestyle = 1
   contrastcolor = GraphColors('gcdata1')
   color = GraphColors('gdata1');
class GraphData2 /
   markersymbol = "plus"
   linestyle = 4
   contrastcolor = GraphColors('gcdata2')
   color = GraphColors('gdata2');
   . . .
```

The style elements displayed in the preceding results are not specified directly in the STATISTICAL style, so they come from the parent DEFAULT style. The `GraphDataDefault` class defines the default marker, marker size, line style, line thickness, marker and line colors, and other colors. The `GraphData1` and `GraphData2` classes define the default appearance of the first two groups. You can convert the hexadecimal colors to decimal as follows:

```
data x;
   input cx $2. (Red Green Blue) (hex2.);
   datalines;
cx2A25D9
cx7C95CA
;

proc print;
run;
```

The results are displayed in Figure 3.15. (Alternatively, you can convert decimal to hex by using PROC PRINT and the statement: `format red green blue hex2.;`)

Figure 3.15 Colors from the Style

Obs	cx	Red	Green	Blue
1	cx	42	37	217
2	cx	124	149	202

You can see that both colors are dominated by the blue component. The first value (the color that is applied to filled areas) is a purer shade of blue; the contrast color (which is applied to markers and lines) has greater contributions from the other colors.

Figure 3.16   Default `GraphDataDefault`

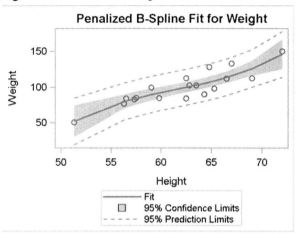

Figure 3.17   Modified `GraphDataDefault`

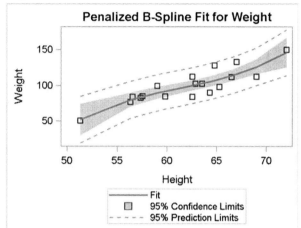

You can make a new style that changes aspects of style elements. For example, the following statements change the `GraphDataDefault` style element:

```
proc template;
   define style Styles.MyStyle;
      parent = Styles.statistical;
      class GraphDataDefault /
         endcolor = GraphColors('gramp3cend')
         neutralcolor = GraphColors('gramp3cneutral')
         startcolor = GraphColors('gramp3cstart')
         markersize = 7px
         markersymbol = "square"
         linethickness = 1px
         linestyle = 1
         contrastcolor = blue
         color = cyan;
      end;
   run;
```

The following steps use the old and new style with the LISTING destination:

```
ods graphics on;
ods listing style=statistical;

proc transreg data=sashelp.class;
   model identity(weight) = pbspline(height);
run;

ods listing style=MyStyle;

proc transreg data=sashelp.class;
   model identity(weight) = pbspline(height);
run;
```

The results are displayed in Figure 3.16 and Figure 3.17. You can see in the plot that the style change affects the marker shapes and colors. Although you can make any style changes that you want, be aware that ad hoc changes such as these might not go well with the other colors and elements in the style. For example, colors might clash and marker and line styles might be duplicated.

3.8 References

Kuhfeld, W. F. (2009), "Modifying ODS Statistical Graphics Templates in SAS 9.2," `http://support.sas.com/rnd/app/papers/modtmplt.pdf`.

Chapter 4

Graph Template Modification

Contents

This chapter explains how to modify the templates that SAS provides.

4.1 Dynamic Variables and Graph Template Modification

This section, along with many other sections throughout the rest of this book, examines some of the graph templates that SAS provides, shows what they have in common and what is different, and shows ways to modify them, thus allowing you to customize the graphs that SAS produces. Understanding how to customize the graphs that SAS produces also helps you learn aspects of the GTL that you can use when writing your own templates. The first few templates that are discussed are chosen because they are small, and they provide complementary insights into template modification. Most templates that SAS provides are more complicated than these.

The default templates supplied by SAS for statistical procedures are often lengthy and complex, because they provide ODS Graphics with comprehensive and detailed information about graph construction. They contain many of the same statements that you saw in the previous examples. However, they often contain statements that do not appear in the previous examples. Some statements are typically not needed when you write your own templates, yet they are important for the templates that SAS provides. This section concentrates on the DYNAMIC statement and also introduces the NOTES and LINEPARM statements.

The rest of this book assumes that the following statements are in effect:

```
ods path (prepend) work.templat(update);
ods graphics on;
ods trace on;
```

The ODS PATH statement stores the modified templates in the SAS WORK library so that they are deleted at the end of your SAS session. The ODS GRAPHICS statement enables ODS Graphics so that graphs are automatically produced by the analytical procedures. The ODS TRACE statement enables ODS trace output so that information about each graph and table (including the graph or table name and the template name) is displayed in the SAS log.

The following template is one of the simplest graph templates that SAS provides for a statistical procedure:

```
define statgraph Stat.MDS.Graphics.Fit;
    notes "MDS Fit Plot";
    dynamic head;
    begingraph / designwidth=defaultdesignheight;
        entrytitle HEAD;
        layout overlayequated / equatetype=square;
            scatterplot y=FITDATA x=FITDIST / markerattrs=(size=5px);
            lineparm slope=1 x=0 y=0 / extend=true lineattrs=GRAPHREFERENCE;
        endlayout;
    endgraph;
end;
```

This template, supplied by SAS for the MDS procedure, creates a scatter plot of two variables, FitData and FitDist, along with a diagonal reference line that passes through the origin. The NOTES statement explains the purpose of the template and does not affect any aspect of the graph. The DYNAMIC statement provides a dynamic variable, Head, that is set by the procedure and used to customize the template at procedure run time. This is analogous to the MVAR and NMVAR statements that are used in previous examples. In previous examples, you specify an MVAR or an NMVAR statement, and then set macro variables for the template to use at procedure run time. In the templates that SAS provides, and with a DYNAMIC statement, the SAS procedure sets dynamic variables for the template to use at procedure run time. If the entry title had been constant, it would have been specified directly in the template. It is not set directly since it can vary depending on the nature of the analysis.

The box that contains the plot is square since the design width is set to the default design height by the option: DESIGNWIDTH=DEFAULTDESIGNHEIGHT on the BEGINGRAPH statement. The LAYOUT OVERLAYEQUATED statement along with the option EQUATETYPE=SQUARE produces a square plot with the axes equated so that a centimeter on one axis represents the same data range as a centimeter on the other axis. The plot title is provided by the run-time evaluation of the dynamic variable Head. It is not unusual for this plot to contain hundreds or even thousands of points, so a five-pixel marker is specified, which is smaller than the seven-pixel marker used by default in most styles.

The LINEPARM statement draws the diagonal reference line, which shows the line of perfect fit. The options X=0 and Y=0 specify coordinates of a point on the line, and the option SLOPE= specifies the slope. The slope of 1 provides a diagonal line. The values of these parameters can be constant, dynamic variables, or macro variables. The option EXTEND=TRUE draws the line to the axes, ignoring any offset area. The LINEATTRS= option specifies the `GraphReference` style element so that the line has the same style as the default reference line style drawn by the REFERENCELINE statement. Notice that the line is drawn after the scatter plot, since its statement comes last. If the statements had been in the reverse order, most or all of the line would have been obscured with large analyses.

The following statements display another of the simpler template definitions—the definition of the scatter plot available in PROC KDE:

```
proc template;
   define statgraph Stat.KDE.Graphics.ScatterPlot;
      dynamic _VAR1NAME _VAR1LABEL _VAR2NAME _VAR2LABEL;
      BeginGraph;
         EntryTitle "Distribution of " _VAR1NAME " by " _VAR2NAME;
         layout Overlay / xaxisopts=(offsetmin=0.05 offsetmax=0.05)
                          yaxisopts=(offsetmin=0.05 offsetmax=0.05);
            ScatterPlot x=X y=Y / markerattrs=GRAPHDATADEFAULT;
         EndLayout;
      EndGraph;
   end;
run;
```

The PROC TEMPLATE and RUN statements have been added to show how you would compile the template if you wanted to modify it.

The DYNAMIC statement specifies four dynamic variables. The dynamic variables _Var1Name and _Var2Name contain the names of the X and Y variables, respectively. The dynamic variables _Var1Label and _Var2Label contain the labels. The names are used in the title. The labels are not used anywhere, but you could modify the template to use them.

There are two ways that you might modify a template. Both are illustrated next. First, you might want to modify the template to use with the procedure for which it is written, PROC KDE. You do so by modifying and compiling the template, and then running the procedure to use the modified template. Alternatively, you might want to use the template outside of the procedure with PROC SGRENDER.

The following steps use PROC KDE and a modified template that uses variable labels instead of the variable names:

```
proc template;
   define statgraph Stat.KDE.Graphics.ScatterPlot;
      dynamic _VAR1NAME _VAR1LABEL _VAR2NAME _VAR2LABEL;
      BeginGraph;
         EntryTitle "Distribution of " _VAR1label " by " _VAR2label;
         layout Overlay / xaxisopts=(offsetmin=0.05 offsetmax=0.05)
                          yaxisopts=(offsetmin=0.05 offsetmax=0.05);
            ScatterPlot x=X y=Y / markerattrs=GRAPHDATADEFAULT;
         EndLayout;
      EndGraph;
   end;
run;
```

Figure 4.1 Plot with Variable Labels

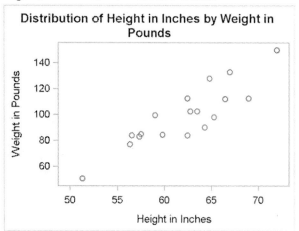

Figure 4.2 PROC SGRENDER's DYNAMIC Statement

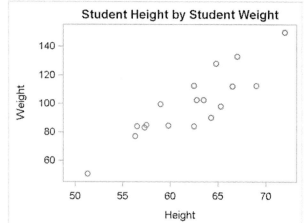

```
proc kde data=sashelp.class;
   bivar height weight / plots=scatter;
   label height = 'Height in Inches' weight = 'Weight in Pounds';
run;
```

The results are displayed in Figure 4.1. This template modification is straightforward. The dynamic variables containing the variable labels are used in the entry title in place of the dynamic variables containing the variable names.

The following steps modify the original template for use outside of PROC KDE but with a data set made from the data object that PROC KDE uses to make the plot:

```
proc template;
   define statgraph Stat.KDE.Graphics.ScatterPlot;
      dynamic _VAR1NAME _VAR1LABEL _VAR2NAME _VAR2LABEL;
      BeginGraph;
         EntryTitle _VAR1NAME " by " _VAR2NAME;
         layout Overlay / xaxisopts=(offsetmin=0.05 offsetmax=0.05)
                          yaxisopts=(offsetmin=0.05 offsetmax=0.05);
            ScatterPlot x=X y=Y / markerattrs=GRAPHDATADEFAULT;
         EndLayout;
      EndGraph;
   end;
run;

ods graphics on;
proc kde data=sashelp.class;
   ods output scatterplot=sp;
   bivar height weight / plots=scatter;
run;

proc sgrender data=sp template=Stat.KDE.Graphics.ScatterPlot;
   dynamic  _VAR1NAME='Student Height' _var2name='Student Weight';
run;
```

Figure 4.3 Modifying Column Names

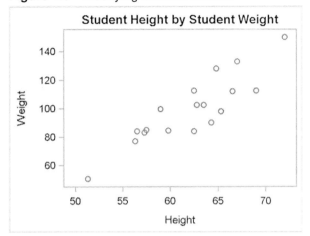

Figure 4.4 Modifying Dynamics

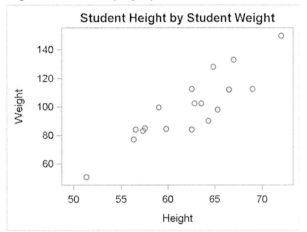

The results are displayed in Figure 4.2. The ODS OUTPUT statement in PROC KDE outputs to a SAS data set the data object that PROC KDE creates and uses to make the plot. Then that data set is input to PROC SGRENDER along with the modified template. The original template could have just as easily been used. This usage of PROC SGRENDER differs from the uses seen previously since this template has dynamic variables. A DYNAMIC statement must be provided with PROC SGRENDER to set the values of the dynamic variables. Note that all of the dynamic variables that are mentioned in the template do not need to be set. Just set the ones that you need. In this case, it is the two variable names. The values of these two dynamic variables appear in the title. The values that are specified on the DYNAMIC statement are not valid SAS variable names, but that does not matter since the dynamic variables are only used to provide title text. You could modify the template for use outside of PROC KDE but with the original Class data set as follows:

```
proc template;
   define statgraph Stat.KDE.Graphics.ScatterPlot;
      dynamic _VAR1NAME _VAR1LABEL _VAR2NAME _VAR2LABEL;
      BeginGraph;
         EntryTitle _VAR1NAME " by " _VAR2NAME;
         layout Overlay / xaxisopts=(offsetmin=0.05 offsetmax=0.05)
                          yaxisopts=(offsetmin=0.05 offsetmax=0.05);
            ScatterPlot x=height y=weight / markerattrs=GRAPHDATADEFAULT;
         EndLayout;
      EndGraph;
   end;
run;

proc sgrender data=sashelp.class template=Stat.KDE.Graphics.ScatterPlot;
   dynamic   _VAR1NAME='Student Height' _var2name='Student Weight';
run;
```

The results are displayed in Figure 4.3. In this step, the SCATTERPLOT statement is modified to specify X=HEIGHT and Y=WEIGHT instead of the original X=X and Y=Y.

The following steps modify the template for use outside of PROC KDE, with the original Sashelp.Class data set, but in a very different way:

```
proc template;
    define statgraph Stat.KDE.Graphics.ScatterPlot;
        dynamic _VAR1NAME _VAR1LABEL _VAR2NAME _VAR2LABEL x y;
        BeginGraph;
            EntryTitle _VAR1NAME " by " _VAR2NAME;
            layout Overlay / xaxisopts=(offsetmin=0.05 offsetmax=0.05)
                             yaxisopts=(offsetmin=0.05 offsetmax=0.05);
                ScatterPlot x=X y=Y / markerattrs=GRAPHDATADEFAULT;
            EndLayout;
        EndGraph;
    end;
run;


proc sgrender data=sashelp.class template=Stat.KDE.Graphics.ScatterPlot;
    dynamic  _VAR1NAME='Student Height' _var2name='Student Weight'
             x='height' y='weight';
run;
```

The results are displayed in Figure 4.4. The original template was written for two data object columns named X and Y. The data set contains two variables named Height and Weight. You can make this template work with the original data set by adding X and Y as dynamic variables in the template definition and then adding x='height' y='weight' to the DYNAMIC statement in the PROC SGRENDER step. This is more complicated than directly changing the X= and Y= specifications in this template. However, if this were a complicated template with the same columns being used multiple places, this would be the easier approach.

Figure 4.5 Default Scree Plot

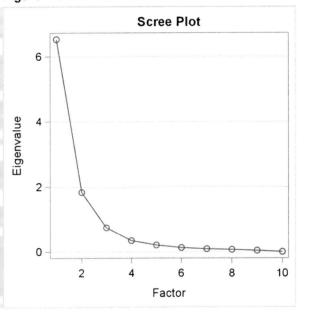

Figure 4.6 Modified Scree Plot

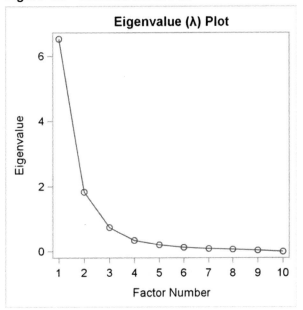

4.2   Changing Titles and Axis Labels

This example shows how to change titles and axis labels in the context of a very simple template. The following step runs PROC FACTOR and produces the eigenvalue (or scree) plot displayed in Figure 4.5:

```
proc factor data=sashelp.cars plots(unpack)=scree;
run;
```

The `plots(unpack)=scree` option produces the scree plot by itself—"unpacked" from its usual location as part of a two-graph panel with the variance-explained plot.

The ODS trace output for the scree plot is as follows:

```
Name:        ScreePlot
Label:       Scree Plot
Template:    Stat.Factor.Graphics.ScreePlot1
Path:        Factor.InitialSolution.ScreeAndVarExp.ScreePlot
```

The following statements display the graph template for the scree plot:

```
proc template;
    source Stat.Factor.Graphics.ScreePlot1;
run;
```

The template source statements are as follows:

```
define statgraph Stat.Factor.Graphics.ScreePlot1;
   notes "Scree Plot for Extracted Eigenvalues";
   BeginGraph / designwidth=DefaultDesignHeight;
      Entrytitle "Scree Plot" / border=false;
      layout overlay / yaxisopts=(label="Eigenvalue" gridDisplay=auto_on)
         xaxisopts=(label="Factor" linearopts=(integer=true));
         seriesplot y=EIGENVALUE x=NUMBER / display=ALL;
      endlayout;
   EndGraph;
end;
```

The DESIGNWIDTH=DEFAULTDESIGNHEIGHT option on the BEGINGRAPH statement specifies that the outer box which contains the graph should be a square whose width is equal to the default graph height. This creates a graph that is designed for a size 480 pixels wide by 480 pixels high. The default design size is 640 pixels wide by 480 pixels high. The ENTRYTITLE statement provides the graph title, in this case "Scree Plot". The BORDER=FALSE option specifies that the title is displayed without a border. In fact, this is the default behavior, so the option is unnecessary. However, it is not unusual to see default specifications in the templates that SAS provides.

The LAYOUT OVERLAY statement provides the label "Eigenvalue" for the vertical or Y axis, provides the label "Factor" for the horizontal or X axis, specifies that grid lines should be produced for the Y axis when the output style favors grids, and specifies that the X axis ticks must be integers. The LINEAROPTS= option is used for options specific to standard axes that depict a linear scaling (as opposed to LOGOPTS=, which is used for log-scale axes).

The graph is a series plot. The Y axis column in the ODS data object is named EigenValue, and the X axis column in the ODS data object is named Number (the factor number). The standard series plot display is a series of lines, but the DISPLAY=ALL option additionally displays the markers (in this case, circles) for the data values.

Notice that the title and the axis labels are all specified directly as literal character strings in this template. You can change any of them and submit the results to SAS. From then on, until you change or delete your custom template in Work.Templat or until you end your SAS session, you will see your customization whenever you run PROC FACTOR.

The following example adds a PROC TEMPLATE statement and a RUN statement, changes the title and an axis label, specifies explicit tick values, and removes the grid and the unnecessary BORDER= option:

```
proc template;
   define statgraph Stat.Factor.Graphics.ScreePlot1;
      notes "Scree Plot for Extracted Eigenvalues";
      BeginGraph / designwidth=DefaultDesignHeight;
         Entrytitle "Eigenvalue ((*ESC*){Unicode Lambda}) Plot";
         layout overlay / yaxisopts=(label="Eigenvalue")
            xaxisopts=(label="Factor Number"
                       linearopts=(tickvaluelist=(1 2 3 4 5 6 7 8 9 10)));
            seriesplot y=EIGENVALUE x=NUMBER / display=ALL;
         endlayout;
      EndGraph;
   end;
run;
```

The title now contains the Greek letter λ, which is specified as an escape sequence followed by a Unicode specification. The tick value list is specified in full because the GTL does not accept standard SAS shorthand lists. The only output from this step is the following log note:

```
NOTE: STATGRAPH 'Stat.Factor.Graphics.ScreePlot1' has been saved to: WORK.TEMPLAT
```

The following step uses the new template to create a scree plot and produces Figure 4.6:

```
proc factor data=sashelp.cars plots(unpack)=scree;
run;
```

The following step restores the default template:

```
proc template;
   delete Stat.Factor.Graphics.ScreePlot1;
run;
```

The only output from this step is the following log note:

```
NOTE: 'Stat.Factor.Graphics.ScreePlot1' has been deleted from: WORK.TEMPLAT
```

Figure 4.7 Default Box Plots

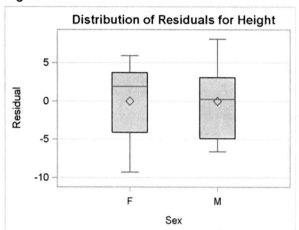

Figure 4.8 Examining Dynamic Variables

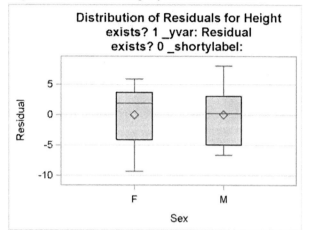

4.3   Changing Titles and Axis Labels Set by Dynamic Variable

In the previous example, the title and both axis labels are specified directly as literal strings in the template. In this example, the procedure provides the title and axis labels at run time. The following step uses the GLIMMIX procedure to create the box plot in Figure 4.7:

```
proc glimmix data=sashelp.class plots=boxplot;
    class sex;
    model height = sex;
run;
```

The trace output for the box plot is as follows:

```
Name:         BoxPlot
Label:        Residuals by Sex
Template:     Stat.Glimmix.Graphics.BoxPlot
Path:         Glimmix.Boxplots.BoxPlot
```

The following statements display the graph template for the box plot:

```
proc template;
    source Stat.Glimmix.Graphics.BoxPlot;
run;
```

The template source statements are as follows:

```
define statgraph Stat.Glimmix.Graphics.BoxPlot;
    dynamic _TITLE _YVAR _SHORTYLABEL;
    BeginGraph;
        entrytitle _TITLE;
        layout overlay / yaxisopts=(gridDisplay=auto_on shortlabel=_SHORTYLABEL)
            xaxisopts=(discreteopts=(tickvaluefitpolicy=rotatethin));
```

```
            boxplot y=_YVAR x=LEVEL / labelfar=on datalabel=OUTLABEL
                primary=true freq=FREQ;
        endlayout;
    EndGraph;
end;
```

The procedure uses dynamic variables to provide text strings and option values to the template. In this case, the dynamic variables are a title, a variable name for the Y axis, and a short variable label for the Y axis. Note that dynamic variables cannot provide any arbitrary syntax. For example, they can provide title text and values of options, but not option names, statement names, layout names, and so on.

Axes can have labels and optionally short labels. The label is displayed if there is sufficient space. Otherwise, the short label is used instead. Axis labels (and short labels) can be specified in the template with a literal string, in the template through a dynamic variable, or implicitly. The axis label comes from the first source that provides a value: the LABEL= option in the template (or the SHORTLABEL= option), the data object column label, or the data object column name.

As a SAS user, you cannot peek into the SAS procedure code to see how the dynamic variables, column names, and column labels are set. However, you can do a bit of detective work to learn about these things. The following steps illustrate one approach:

```
proc template;
    define statgraph Stat.Glimmix.Graphics.BoxPlot;
        dynamic _TITLE _YVAR _SHORTYLABEL;
        BeginGraph;
            entrytitle _TITLE;
            entrytitle "exists? " eval(exists(_yvar)) " _yvar: " _yvar;
            entrytitle "exists? " eval(exists(_shortylabel))
                    " _shortylabel: " _shortylabel;
            layout overlay / yaxisopts=(gridDisplay=auto_on shortlabel=_SHORTYLABEL)
                xaxisopts=(discreteopts=(tickvaluefitpolicy=rotatethin));
                boxplot y=_YVAR x=LEVEL / labelfar=on datalabel=OUTLABEL
                    primary=true freq=FREQ;
            endlayout;
        EndGraph;
    end;

proc glimmix data=sashelp.class plots=boxplot;
    class sex;
    ods output boxplot=bp;
    model height = sex;
run;

proc contents p;
    ods select position;
run;
```

The graph is displayed in Figure 4.8, and the data object contents are displayed in Figure 4.9. The first title is unmodified and simply displays the value of the _Title dynamic variable. Following that, this template is temporarily modified by adding two new ENTRYTITLE statements to report on both the existence and the value of two of the dynamic variables. The expression **eval(exists(**_dynamic-variable_**))** resolves to 1 (for true) when the dynamic variable is set by the procedure and 0 (for false) when it is not set. It is not unusual for a procedure to conditionally set dynamic variables. A specification of *option=dynamic* is ignored when the dynamic variable does not exist. After the existence information is displayed, the value (if any) is displayed.

Figure 4.9 Contents of a Data Object

```
                      The CONTENTS Procedure

                   Variables in Creation Order

   #    Variable                          Type   Len   Format   Label

   1    BOX__YVAR_X_LEVEL_DATALABEL_O__Y   Num     8             Residual
   2    BOX__YVAR_X_LEVEL_DATALABEL_O_ST   Char   10
   3    BOX__YVAR_X_LEVEL_DATALABEL_O__X   Char    1             Sex
   4    BOX__YVAR_X_LEVEL_DATALABEL_O_DL   Num     8    BEST8.   Index
   5    Residual                          Num     8
   6    Level                             Char    1             Sex
   7    OutLabel                          Num     8    BEST8.   Index
```

Figure 4.8 shows that the Y axis column is Residual and the short Y axis label is undefined. The PROC CONTENTS information confirms that the data object has a column called Residual for the Y axis and a column called Level with a label of "Sex" for the X axis. The Y axis column name and the X axis column label become the axis labels. The contents information also displays other columns in the data object.[1]

There are a number of ways that you can modify templates beyond simply adding or replacing text. For example, you can use the dynamic variables that are provided in creative ways, such as using the title as a label for the Y axis: `yaxisopts=(label=_title . . .)`.

---

[1]Data objects come in many varied forms. You should not expect them to be pretty or well organized for display or subsequent processing. Although you can process them in any way you choose, they are designed for input to one or more templates and very little else. On some occasions, extra columns or extra dynamic variables might be created but not used. These represent cases where the procedure writer recognized possibilities for future processing and tried to facilitate them. They might be helpful when they occur, but most data objects or templates do not have such information. This data object has a number of manufactured and verbose names. You often see names like these when the values that are plotted are statistics of some sort or are computed by ODS Graphics.

Figure 4.10 Modified Box Plots

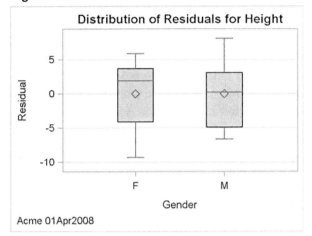

Figure 4.11 Footnote Added with a Macro

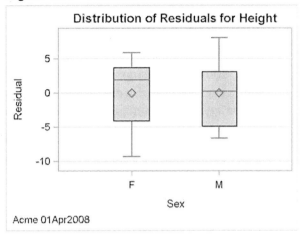

This next example will instead replace the X axis label and add a footnote (horizontally aligned on the left using a font that is appropriate for footnotes or second title lines), both through macro variables, as follows:

```
proc template;
   define statgraph Stat.Glimmix.Graphics.BoxPlot;
      dynamic _TITLE _YVAR _SHORTYLABEL;
      mvar datetag xlabel;
      BeginGraph;
         entrytitle _TITLE;
         entryfootnote halign=left textattrs=graphvaluetext datetag;
         layout overlay / yaxisopts=(gridDisplay=auto_on shortlabel=_SHORTYLABEL)
            xaxisopts=(label=xlabel
                        discreteopts=(tickvaluefitpolicy=rotatethin));
            boxplot y=_YVAR x=LEVEL / labelfar=on datalabel=OUTLABEL
               primary=true freq=FREQ;
         endlayout;
      EndGraph;
   end;
run;

%let DateTag = Acme 01Apr2008;
%let xlabel  = Gender;

proc glimmix data=sashelp.class plots=boxplot;
   class sex;
   ods output boxplot=bp;
   model height = sex;
run;
```

The new graph is displayed in Figure 4.10.

In this example, there is a new footnote, which comes from the value of the macro variable DateTag. DateTag is specified in the MVAR (macro variable) statement. When the MVAR statement is used, the template is compiled and the value of the macro variable is substituted when the template is used by the procedure. This approach lets you modify and compile the template once and then use it repeatedly with different values of the macro variable without ever having to recompile the template.

You usually use this approach when you make persistent changes in Sasuser.Templat or some other permanent item store. An alternate approach is to use the following statement without using an MVAR statement:

```
entryfootnote "&datetag";
```

In this approach, the template is compiled and the value of the macro variable is substituted at compile time. The value cannot change in this approach unless you recompile the template. In this case, the approach does not matter because the template is compiled and immediately used.

The X axis label is set to "Gender" using the macro variable xlabel. The X axis change is ad hoc, so changes such as this are usually made temporarily.

The following step restores the default template:

```
proc template;
    delete Stat.Glimmix.Graphics.BoxPlot;
run;
```

If your only goal is to add or change a footnote or title, there is an easier, alternative mechanism. SAS provides a new autocall macro, **ModTmplt**, that you can use for this purpose (Kuhfeld 2009). This macro is used in the following example:

```
title;
footnote 'halign=left textattrs=graphvaluetext "Acme 01Apr2008"';
%modtmplt (template=Stat.Glimmix.Graphics.BoxPlot, steps=t,
          options=titles noquotes)
footnote;

proc glimmix data=sashelp.class plots=boxplot;
    class sex;
    ods output boxplot=bp;
    model height = sex;
run;

%modtmplt (template=Stat.Glimmix.Graphics.BoxPlot, steps=d)
```

The TITLE statement clears the default title of "The SAS System". The FOOTNOTE statement provides the footnote along with options to place the footnote on the left using the font that is used for values. The **ModTmplt** macro modifies the box plot template. Only one macro step is run: the template modification step (STEPS=T). OPTIONS=TITLES adds SAS system titles and footnotes (those specified in TITLE and FOOTNOTE statements) to the existing graph titles and footnotes. OPTIONS=NOQUOTES moves the footnote from the FOOTNOTE statement to the ENTRYFOOTNOTE statement but without the outer quotes. You must specify this option if you want to specify ENTRYTITLE or ENTRYFOOTNOTE options in your TITLE or FOOTNOTE statement. The next FOOTNOTE statement clears the footnote so that it affects only the box plot template and does not otherwise affect the analysis. PROC GLIMMIX makes the plot. The final call to the macro deletes the modified template (STEPS=D). The results are displayed in Figure 4.11.

Figure 4.12   Transposed Scatter Plot with Labels

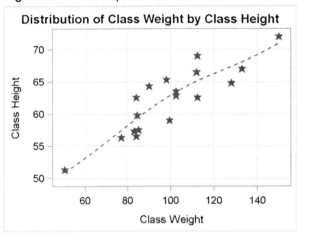

Figure 4.13   Scatterplot with Numerous Modifications

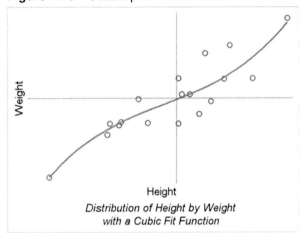

4.4   Modifying Colors, Lines, Markers, Axes, and Reference Lines

This example shows how to change the axes and the inside of the plot including the lines and markers. The following template source appeared in the section "Dynamic Variables and Graph Template Modification" on page 153:

```
define statgraph Stat.KDE.Graphics.ScatterPlot;
   dynamic _VAR1NAME _VAR1LABEL _VAR2NAME _VAR2LABEL;
   BeginGraph;
      EntryTitle "Distribution of " _VAR1NAME " by " _VAR2NAME;
      layout Overlay / xaxisopts=(offsetmin=0.05 offsetmax=0.05)
         yaxisopts=(offsetmin=0.05 offsetmax=0.05);
         ScatterPlot x=X y=Y / markerattrs=GRAPHDATADEFAULT;
      EndLayout;
   EndGraph;
end;
```

The entry title is a mix of constant text and dynamic variables that provide variable names. The procedure writer has provided you with additional dynamic variables that provide the variable labels. This template also has offset options specified. These options are frequently used in scatter plots and other graphs. They add a small amount of white space to the left or bottom (OFFSETMIN=) and to the right or top (OFFSETMAX=) of the specified axis. The SCATTERPLOT statement has a MARKERATTRS= option, which references the **GraphDataDefault** style element. This style element also controls line color and thickness.

The following statements switch the Y and X axis variables, use variable labels instead of variable names in the title, change the marker characteristics, add a nonlinear penalized B-spline fit function, and add grids:

```
proc template;
    define statgraph Stat.KDE.Graphics.ScatterPlot;
        dynamic _VAR1NAME _VAR1LABEL _VAR2NAME _VAR2LABEL;
        BeginGraph;
            EntryTitle "Distribution of " _VAR2LABEL " by " _VAR1LABEL;
            layout Overlay /
                xaxisopts=(offsetmin=0.05 offsetmax=0.05 griddisplay=on)
                yaxisopts=(offsetmin=0.05 offsetmax=0.05 griddisplay=on);
                pbsplineplot x=y y=x / lineattrs=(color=red pattern=2 thickness=1);
                ScatterPlot  x=y y=x / markerattrs=(color=green size=5px
                                            symbol=starfilled weight=bold);
            EndLayout;
        EndGraph;
    end;
run;


proc kde data=sashelp.class;
    label height = 'Class Height' weight = 'Class Weight';
    bivar height weight / plots=scatter;
run;
```

The results are displayed in Figure 4.12.

Note that the addition of the variable labels in the PROC KDE step also changes the axis labels because no axis labels are explicitly specified in the template. The PBSPLINEPLOT statement includes the LINEATTRS= option which specifies the color (red), pattern (2, dashed), and thickness (1 pixel) of the fit function. The SCATTERPLOT statement includes the MARKERATTRS= option which specifies the color (green), size (5 pixels), symbol (a filled star), and weight (bold) of the marker or symbol.

The starting point for the next step is the original PROC KDE scatter plot template, rather than the modified template. The goal in this example is to produce the scatter plot displayed in Figure 4.13 with axes passing through the center of the data. The following steps make a highly modified scatter plot:

```
proc template;
    define statgraph Stat.KDE.Graphics.ScatterPlot;
        dynamic _VAR1NAME _VAR1LABEL _VAR2NAME _VAR2LABEL;
        BeginGraph;
            EntryFootNote "Distribution of " _VAR1NAME " by " _VAR2NAME;
            EntryFootNote "with a Cubic Fit Function";
            layout Overlay / walldisplay=none
                xaxisopts=(display=(label))
                yaxisopts=(display=(label));
                referenceline y=eval(mean(y));
                referenceline x=eval(mean(x));
                ScatterPlot x=X y=Y / markerattrs=GRAPHDATADEFAULT;
                regressionplot x=x y=y / degree=3;
            EndLayout;
        EndGraph;
    end;
run;
```

```
proc kde data=sashelp.class;
   bivar height weight / plots=scatter;
run;
```

The template is modified by adding an MVAR statement to use the mean height and mean weight. Specifically, a reference line is displayed at the mean for each axis. The title is changed to a footnote, and a second footnote is added. The LAYOUT OVERLAY statement now has a WALLDISPLAY=NONE option to suppress the axes, and only the labels are displayed on each axis. A cubic-polynomial fit function is also added. The results are displayed in Figure 4.13.

The LAYOUT OVERLAY block has four statements in it. The statements are executed in the order in which they are specified in the LAYOUT OVERLAY block. Reference lines are displayed first. Therefore any point or function that coincides with the reference line is displayed on top of the reference line. Similarly, the fit function is displayed rather than the points in the places where they coincide. You can vary the order of the statements if you prefer some other effect. The plot has no axes, no ticks, no tick labels, and no wall. (The wall is the area inside the plot axes, which can be a different color from the background color outside of the axes.) Instead, the plot simply has reference lines at the means and axis labels. Many variations can be tried. In the interest of space, several variations are discussed but their results are not shown.

The following statement displays a left axis and a bottom axis (but no top axis or right axis), and the color outside the axes matches the color inside:

```
layout Overlay / walldisplay=none;
```

The following statement displays all four axes, and the color outside the axes matches the color inside:

```
layout Overlay / walldisplay=(outline);
```

The following statement suppresses all axis information (the axes, the ticks, the tick labels, the axis labels, and the wall):

```
layout Overlay / walldisplay=none
   xaxisopts=(display=none) yaxisopts=(display=none);
```

Figure 4.14 Default Residual Histogram

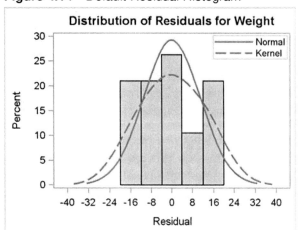

Figure 4.15 Residual Histogram with Repositioned Legend

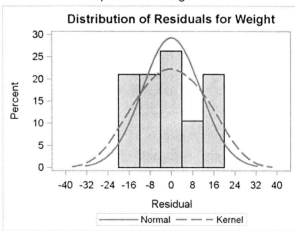

4.5 Legends

This section creates a plot with a legend. The following step runs the GLM procedure and produces a residual histogram:

```
proc glm plots=diagnostics(unpack) data=sashelp.class;
   model weight = height;
   ods output residualhistogram=hr;
run;

proc contents p;
   ods select position;
run;
```

This type of graph (shown in Figure 4.14) is used in many procedures.

The trace output for the residual histogram is as follows:

```
Name:        ResidualHistogram
Label:       Residual Histogram
Template:    Stat.GLM.Graphics.ResidualHistogram
Path:        GLM.ANOVA.Weight.DiagnosticPlots.ResidualHistogram
```

The following statements display the graph template for the residual histogram:

```
proc template;
   source Stat.GLM.Graphics.ResidualHistogram;
run;
```

The template source statements are as follows:

```
define statgraph Stat.GLM.Graphics.ResidualHistogram;
   notes "Residual Histogram with Overlayed Normal and Kernel";
   dynamic Residual _DEPNAME;
   BeginGraph;
      entrytitle "Distribution of Residuals" " for " _DEPNAME;
      layout overlay / xaxisopts=(label="Residual")
         yaxisopts=(label="Percent");
         histogram RESIDUAL / primary=true;
         densityplot RESIDUAL / name="Normal" legendlabel="Normal"
            lineattrs=GRAPHFIT;
         densityplot RESIDUAL / kernel () name="Kernel" legendlabel="Kernel"
            lineattrs=GRAPHFIT2;
         discretelegend "Normal" "Kernel" / across=1 location=inside
            autoalign=(topright topleft top);
      endlayout;
   EndGraph;
end;
```

This graph template creates a histogram of residuals. On top of the histogram is a normal density plot, and on top of both is a kernel density plot. Additionally, a legend is positioned inside the graph. The preferred position is in the top right, but ODS Graphics automatically repositions the legend in the top left or top center if there are conflicts between the legend and the histogram or functions in the top right.

The ENTRYTITLE statement specifies the title, which consists of literal text and a dynamic variable that contains the dependent variable name. The LAYOUT OVERLAY statement specifies the labels for both axes. Since the labels never vary in this template, they are specified directly in the template. The HISTOGRAM statement creates a histogram from the data object column named Residual. It is the primary statement in the overlay. The data columns from the primary statement determine the default axis types and default axis labels. By default, the first graph statement is the primary statement. Hence, in this case the PRIMARY= option is not needed. You must specify PRIMARY=TRUE when you want a statement other than the first to control the axes. The color, width, and line style for the normal density plot comes from the `GraphFit` style element (blue and solid in this style), and for the kernel density plot comes from the `GraphFit2` style element (red and dashed in this style). All graphs are based on the same data object column, Residual, and the kernel density plot uses default options for finding the kernel density.

The contents of the data object are displayed in Figure 4.16. From the original input variable Residual, six other variables are created by the HISTOGRAM and the two DENSITYPLOT statements. The X and Y axis variables for the density plot are BIN_RESIDUAL___X and BIN_RESIDUAL___Y; for the normal density plot they are NORMAL_RESIDUAL___X and NORMAL_RESIDUAL___Y; and for the kernel density plot they are KERNEL_RESIDUAL___X and KERNEL_RESIDUAL___Y. If you display the output data set created from this data object, you will see that the variables do not have the same number of nonmissing values. Some, such as the histogram values, have many fewer than the others. In this case, the computed density values have many more values than the raw residuals. Data objects are often constructed from pieces of very different sizes.

Figure 4.16 Contents of the Residual Histogram Data Object

```
              The CONTENTS Procedure

            Variables in Creation Order

     #     Variable             Type    Len    Label

     1     Dependent            Char     8
     2     BIN_RESIDUAL___X     Num      8     Residual
     3     BIN_RESIDUAL___Y     Num      8     Percent
     4     NORMAL_RESIDUAL__X   Num      8     Residual
     5     NORMAL_RESIDUAL__Y   Num      8     Percent
     6     KERNEL_RESIDUAL__X   Num      8     Residual
     7     KERNEL_RESIDUAL__Y   Num      8     Percent
     8     Residual             Num      8
```

All of the remaining options concern the legend. The discrete legend is produced by the DIS-CRETELEGEND statement. In contrast, a continuous legend is used to produce a color "thermometer" legend when point or surface colors vary continuously as a function of a third variable (see the section "Continuous Legend" on page 72). The legend is constructed from the statements named "Normal" and "Kernel" by the NAME= option in each of the two DENSITYPLOT statements. The labels for these two legend components come from the LEGENDLABEL= options. The legend has only one component in each row due to the ACROSS=1 option.

The following steps modify the graph by moving the legend outside the graph and by removing the ACROSS= option, which for this graph produces a legend with one row and two entries:

```
proc template;
   define statgraph Stat.GLM.Graphics.ResidualHistogram;
      notes "Residual Histogram with Overlayed Normal and Kernel";
      dynamic Residual _DEPNAME;
      BeginGraph;
         entrytitle "Distribution of Residuals" " for " _DEPNAME;
         layout overlay / xaxisopts=(label="Residual")
            yaxisopts=(label="Percent");
            histogram RESIDUAL / primary=true;
            densityplot RESIDUAL / name="Normal"
               legendlabel="Normal" lineattrs=GRAPHFIT;
            densityplot RESIDUAL / kernel ()
               name="Kernel" legendlabel="Kernel" lineattrs=GRAPHFIT2;
            discretelegend "Normal" "Kernel";
         endlayout;
      EndGraph;
   end;
run;

proc glm plots=diagnostics(unpack) data=sashelp.class;
   model weight = height;
run;
```

The results are displayed in Figure 4.15.

4.6 Lattice Layout and Panels

The templates that SAS provides have so far been simple in that they produce a graph with one panel. Those templates consist of a single LAYOUT OVERLAY block with one or more plotting statements inside. However, many graphs consist of two or more panels within a single display. For example, the scree plot displayed in Figure 4.5 is, by default, part of a two-panel display. It is produced when you run PROC FACTOR without the UNPACK option as follows:

```
proc factor data=sashelp.cars plots=scree;
run;
```

The graph is displayed in Figure 4.17.

The trace output (not shown) shows that the template is called `Stat.Factor.Graphics.ScreePlot2`. A slight simplification of the template source statements is as follows:

```
define statgraph Stat.Factor.Graphics.ScreePlot2;
    notes "Scree and Proportion Variance Explained Plots";
    BeginGraph / DesignHeight=360px;
        layout lattice / rows=1 columns=2 columngutter=30;
            layout overlay / yaxisopts=(label="Eigenvalue" gridDisplay=auto_on)
                xaxisopts=(label="Factor" linearopts=(integer=true));
                entry "Scree Plot" / textattrs=GRAPHLABELTEXT location=outside;
                seriesplot y=EIGENVALUE x=NUMBER / display=ALL;
            endlayout;
            layout overlay / yaxisopts=(label="Proportion" gridDisplay=auto_on)
                xaxisopts=(label="Factor" linearopts=(integer=true));
                entry "Variance Explained" / textattrs=GRAPHLABELTEXT
                                              location=outside;
                seriesplot y=PROPORTION x=NUMBER / display=ALL legendlabel=
                    "Proportion" name="Proportion";
                seriesplot y=CUMULATIVE x=NUMBER / lineattrs=GRAPHDATADEFAULT (
                    pattern=dot) display=ALL LegendLabel="Cumulative" name=
                    "Cumulative" primary=true;
                DiscreteLegend "Cumulative" "Proportion" / across=1 border=1;
            endlayout;
        endlayout;
    EndGraph;
end;
```

The template begins with a BEGINGRAPH statement. Most templates do not contain a specific numerical size for the overall graph area. This template does, so that the two resulting plots are approximately square. At the default size, the plots are tall and thin. Note that size is a "design height" rather than a hardcoded size. The graph is designed at a height of 360 pixels, but it can be stretched or shrunk to other sizes while preserving the aspect ratio.

The next layer is a LAYOUT LATTICE block that creates a display with one row and two columns. Row and column gutters are frequently specified in lattice layouts. The COLUMNGUTTER=30 option ensures that there are 30 pixels between the two columns of graphs. (By default, the plots are closer than that.) Inside of the lattice layout are two LAYOUT OVERLAY blocks, one for each graph. Each individual LAYOUT OVERLAY block is designed in much the same way it would be designed if it were in a one-panel display. However, in practice it is not unusual for an "unpacked graph" (a graph produced in a single panel) to be different from the same graph packed into a display with other graphs.

Figure 4.17 Panel with Two Graphs

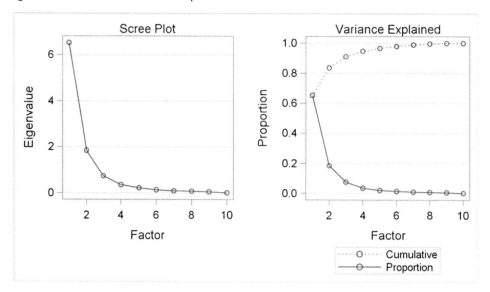

In some cases, the unpacked plot might have additional graph elements due to the increased room in the unpacked plot. One difference in the paneled plot is the title. In this case, the goal is to have two titles, one for each plot with no overall title. Hence, there is no ENTRYTITLE statement. Instead, there are two ENTRY statements, which place the title outside (and above) each plot by using the `GraphLabelText` style element. By default, without this style specification, the text would be smaller and would not look like other titles. The LOCATION=OUTSIDE option is an example of one of the undocumented options that are specified in some templates.

The last SERIESPLOT statement specifies the `lineattrs=GraphDataDefault(pattern=dot)` option. The properties of the line produced by this statement are controlled by the `GraphDataDefault` style element. However, one aspect of the style (namely the line pattern) is overridden and a dotted line is used instead. PATTERN= is one of the options in the LINEATTRS= option, rather than the name of a style element (which is `MarkerSymbol`). Since `GraphDataDefault` is the default style for the first series plot in the second overlay layout, the specification in the second series plot ensures that the two series have identical styles (except for one aspect) so that they can be distinguished in the legend. Most SAS/STAT templates do not hardcode graph elements such as this (the dotted line); usually they strictly use style elements. However, on occasion you will see explicit specifications. For example, the LOESS and TRANSREG procedures use the specification `MarkerAttrs=GraphData1(symbol=star size=15)` to mark the minimum of a function that is being optimized. This specification produces a large star. Numerical sizes like SIZE=15 are design sizes; they can be stretched or shrunk to other sizes while preserving the aspect ratio.

You can do many things to modify this template. You can change the titles, axis labels, colors, markers, and so on. All of these are illustrated in other parts of this book. You can switch the order of the layouts and put the variance-explained plot first; you can provide an overall title; and so on. However, rather than perform familiar or obvious changes, the rest of the book concentrates on understanding other aspects of the GTL and the complex templates that you might encounter.

4.7   Conditional Template Logic

This example explains the layout of a template with conditional logic and nested IF statements. You might need to understand conditional template logic when you determine which part of a template to modify. The survival estimate plot from the LIFETEST procedure has a long and complicated template, which includes nested IF statements. We assume at this point that you know how to run the procedure and find the name of the template. A very small part of the template is as follows:

```
define statgraph Stat.Lifetest.Graphics.ProductLimitSurvival;
   dynamic . . .;
   BeginGraph;
      if (NSTRATA=1)
         if (EXISTS(STRATUMID))
            entrytitle "Product-Limit Survival Estimate" " for " STRATUMID;
      else
            entrytitle "Product-Limit Survival Estimate";
      endif;
      if (PLOTATRISK)
            entrytitle "with Number of Subjects at Risk" / textattrs=
            GRAPHVALUETEXT;
      endif;
      layout overlay . . .;
            . . .
      endlayout;
      else
            entrytitle "Product-Limit Survival Estimates";
      if (EXISTS(SECONDTITLE))
            entrytitle SECONDTITLE / textattrs=GRAPHVALUETEXT;
      endif;
      layout overlay . . .;
            . . .
      endlayout;
      endif;
   EndGraph;
end;
```

This layout is confusing even when displayed like this with most details removed. In its original form at 145 lines, it is even more confusing. To understand the layout of this template, you must carefully evaluate the IF, ELSE, ENDIF, structure. The following step shows the structure manually re-indented, with additional details removed and additional white space added:

```
define statgraph Stat.Lifetest.Graphics.ProductLimitSurvival;
    dynamic . . .;
    BeginGraph;
        if (NSTRATA=1)

            if (EXISTS(STRATUMID)) entrytitle . . .;
            else entrytitle . . .;
            endif;

            if (PLOTATRISK) entrytitle . . .;
            endif;

            layout overlay ...;
                . . .
            endlayout;

        else

            entrytitle . . .;

            if (EXISTS(SECONDTITLE)) entrytitle . . .;
            endif;

            layout overlay . . .;
                . . .
            endlayout;

        endif;
    EndGraph;
end;
```

The IF and ELSE statements do not perform as do the similarly named statements in the DATA step. There are no DO and END statements. When the first IF condition is true, the statements under the first IF statement are executed until control reaches the ELSE statement at the same indentation level. The statements in the ELSE block include everything below the ELSE and through the ENDIF statement at the same level. Inside the first IF block, a title is provided if a condition is true. Otherwise a different title is provided, and that block ends with the first ENDIF statement.

If there is one stratum (NSTRATA=1), then the graph consists of one of two conditional titles followed by a conditional second title followed by a graph defined in a layout. With more than one stratum, the graph consists of an unconditional title, a conditional second title, and a graph defined in a different layout. Sometimes the easiest way to understand a template structure is to do precisely what is done here: copy the template and remove details until you are left with an outline of the overall structure. Then use that knowledge to go back and evaluate and modify the full template.

The following provides a similar manual re-edit and pruning, this time concentrating on the titles:

```
define statgraph Stat.Lifetest.Graphics.ProductLimitSurvival;
   dynamic . . .;
   BeginGraph;
      if (NSTRATA=1)
         if (EXISTS(STRATUMID))
            entrytitle "Product-Limit Survival Estimate" " for " STRATUMID;
         else
            entrytitle "Product-Limit Survival Estimate";
         endif;
         if (PLOTATRISK)
            entrytitle "with Number of Subjects at Risk" / textattrs=GRAPHVALUETEXT;
         endif;
         . . .
      else
         entrytitle "Product-Limit Survival Estimates";
         if (EXISTS(SECONDTITLE))
            entrytitle SECONDTITLE / textattrs=GRAPHVALUETEXT;
         endif;
         . . .
      endif;
   EndGraph;
end;
```

You can see that titles can come from literal strings, dynamic variables, or both. If you are unclear about which title appears in the output, you can temporarily change the titles by adding some text to clearly show which is which. The following statements show how:

```
entrytitle "(1) Product-Limit Survival Estimate" " for " STRATUMID;
entrytitle "(2) Product-Limit Survival Estimate";
entrytitle "(3) with Number of Subjects at Risk" / textattrs=GRAPHVALUETEXT;
entrytitle "(4) Product-Limit Survival Estimates";
entrytitle "(5)" SECONDTITLE / textattrs=GRAPHVALUETEXT;
```

If you submit the full template with titles like these, you can clearly see whether a title is used and where it is used. You can apply the same technique to axis labels, legend labels, and any other text in the template. Then you can remove the identification numbers, modify the text of interest, and submit the modified template. For example, you might wish to change the first, second, and fourth title to "Kaplan-Meier Plot". Note that it is not always sufficient to find and change the first entry title in a template.

In a previous example, an ENTRY statement specified the style element **GraphLabelText** so that the entry text would look like a title. In this template, an ENTRYTITLE statement specifies the style element **GraphValueText** so that second titles are subordinate (less bold or smaller according to the style) to the first title lines.

IF, ELSE, and ENDIF statements cannot be used in arbitrary ways. The GTL code that is conditional must be complete. For example, the following statements produce an error:

```
if ( exists(SQUAREPLOT) )                            /* Wrong! */
   layout overlayequated / equatetype=square;  /* Wrong! */
 else                                                /* Wrong! */
   layout overlay;                                   /* Wrong! */
 endif;                                              /* Wrong! */
     scatterplot x=XVAR y=YVAR;                      /* Wrong! */
 endlayout;                                          /* Wrong! */
```

The following statements are correct:

```
if (exists(SQUAREPLOT))
   layout overlayequated / equatetype=square;
      scatterplot x=XVAR y=YVAR;
   endlayout;
else
   layout overlay;
      scatterplot x=XVAR y=YVAR;
   endlayout;
endif;
```

The incorrect example attempts to conditionally execute a complete statement, but only complete layouts (not merely complete layout statements) can be conditionally executed. Also note that IF conditions determine what is rendered in the plot rather than what is computed for the data object. For example, the following step attempts to compute a LOESS fit whether or not the LOESSPLOT dynamic variable is defined:

```
if (exists(LOESSPLOT))
   loessplot y=LOESS x=X;
endif;
```

Since the LOESS fit is computationally expensive, procedure writers use a different approach to conditionally compute results only when needed. If either LOESS or X is a dynamic variable that is not defined to a data object column name, then the computation is not performed.

Figure 4.18 Diagnostics Panel for the REG Procedure

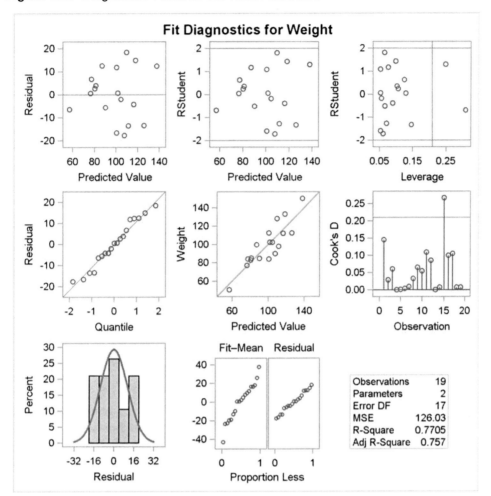

4.8 Paneled Displays

This example discusses how to understand the overall structure of a paneled display with many graphs and how to isolate individual graphs that you might want to modify. The following step fits a regression model and displays a set of model fit diagnostics:

```
ods graphics on;

proc reg data=sashelp.class;
   model weight = height;
run; quit;
```

The diagnostics panel is displayed in Figure 4.18.

The rendered version of the diagnostics panel template, `Stat.Reg.Graphics.DiagnosticsPanel`, has 271 lines. A very small portion of it is as follows:

```
define statgraph Stat.Reg.Graphics.DiagnosticsPanel;
   notes "Diagnostics Panel";
   dynamic . . .;
   BeginGraph / designheight=defaultDesignWidth;
      entrytitle . . .;
      layout lattice / columns=3 rowgutter=10 columngutter=10
         shrinkfonts=true rows=3;
         layout overlay . . . scatterplot . . . endlayout;
         layout overlay . . . scatterplot . . . endlayout;
         layout overlay . . . scatterplot . . . endlayout;
         layout overlay . . . scatterplot . . . endlayout;
         layout overlayequated . . . scatterplot . . . endlayout;
         layout overlay . . . needleplot . . . endlayout;
         layout overlay . . . histogram . . . densityplot . . . endlayout;
         layout lattice / columns=2 rows=1 rowdatarange=unionall columngutter=0
            . . .
            layout overlay . . . scatterplot . . . endlayout;
            layout overlay . . . scatterplot . . . endlayout;
            . . .
         endlayout;
         . . .
         layout overlay;
            layout gridded / columns=_NSTATSCOLS valign=center border=TRUE
               BackgroundColor=GraphWalls:Color Opaque=true;
               . . . entry halign=left "Observations" / valign=top;
               . . . entry halign=right eval (PUT(_NOBS,BEST6.)) / valign=top;
               . . .
            endLayout;
         endif;
      . . .
      endlayout;
   EndGraph;
end;
```

The paneled display is large and square (although greatly reduced from the default size for this book) and is designed with a height equal to the default width. It has a single overall title for the display. It consists of a 3 by 3 lattice of nine entries. The first eight panels are graphs, and the last is a table of statistics. The graphs that are defined in the overlay layouts fill in the display in order from left to right and from top to bottom. The first four graphs are ordinary scatter plots; the fifth is an equated scatter plot where both axes are equated to represent the same data range; the sixth is a needle plot; the seventh is an overlay of a histogram and a density plot; the eighth is another lattice, this one consisting of two scatter plots; and the ninth and final panel in the outer lattice is a grid that contains statistic names and their values. The outermost lattice specifies the SHRINKFONTS=TRUE option. This option is commonly specified in outer lattices and specifies that fonts can be scaled down when the graph is reduced in size. Without this option, the text is typically too large in reduced versions of displays such as Figure 4.18.

Even when the overall template is huge, you can often find and isolate small template components that are easily understood. For example, the first plot, which displays residuals and predicted values, is created from the following layout:

```
layout overlay / xaxisopts=(shortlabel='Predicted');
   referenceline y=0;
   scatterplot y=RESIDUAL x=PREDICTEDVALUE / primary=true datalabel=
      _OUTLEVLABEL rolename=(_tip1=OBSERVATION _id1=ID1 _id2=ID2 _id3=
      ID3 _id4=ID4 _id5=ID5) tip=(y x _tip1 _id1 _id2 _id3 _id4 _id5);
endlayout;
```

The DATALABEL= option provides labels for the markers when the dynamic variable _OutLevLabel exists. The ROLENAME= and TIP= options create tooltips in HTML. Tooltips are text boxes that appear in HTML output when your mouse pointer hovers over a part of the plot. Tips are produced for the Y axis column, the X axis column, and additional columns _tip1 and _id1 through _id5. The columns x and y have predefined roles as axis variables. In contrast, the other tips are provided for columns that are identified through the ROLENAME= option. You must provide role names for columns that do not have automatic roles (such as the axis columns) and use the role names rather than the column names in the TIP= option. You can modify the tooltips by adding, deleting, or changing columns specified in these lists. These options usually are specified in templates for graphs that display data or computed values with a one-to-one correspondence with the data (for example, independent variable, dependent variable, predicted values, residuals, leverage, and variables specified in the procedure's ID statement).

Part of the gridded layout that composes the ninth panel is as follows (after some manual indentation adjustments):

```
if (_SHOWNOBS^=0)
   entry halign=left "Observations" / valign=top;
   entry halign=right eval (PUT(_NOBS,BEST6.)) / valign=top;
endif;
if (_SHOWTOTFREQ^=0)
   entry halign=left "Total Frequency" / valign=top;
   entry halign=right eval (PUT(_TOTFREQ,BEST6.)) / valign=top;
endif;
if (_SHOWNPARM^=0)
   entry halign=left "Parameters" / valign=top;
   entry halign=right eval (PUT(_NPARM,BEST6.)) / valign=top;
endif;
```

Do not rely on the indentation provided by PROC TEMPLATE and the SOURCE statement to see the structure of a template. Re-indent the template yourself to make it clearer. Each statistic is added to the display conditional on a dynamic variable. First, a label is displayed on the left followed by a value on the right. In a table such as this, you could change the labels, change the formats, remove statistics, or reorder them.

The layout for the fourth graph, the normal quantile plot of the residuals, which is displayed in the second row and first column of the panel, is as follows:

```
layout overlay / yaxisopts=(label="Residual" shortlabel="Resid")
   xaxisopts=(label="Quantile");
   lineparm slope=eval (STDDEV(RESIDUAL)) y=eval (MEAN(RESIDUAL)) x=0
      / extend=true lineattrs=GRAPHREFERENCE;
   scatterplot y=eval (SORT(DROPMISSING(RESIDUAL))) x=eval (
      PROBIT((NUMERATE(SORT(DROPMISSING(RESIDUAL))) -0.375)/
      (0.25 + N(RESIDUAL)))) / markerattrs=GRAPHDATADEFAULT
      primary=true rolename=(s=eval (SORT(DROPMISSING(RESIDUAL)))
      nq=eval (PROBIT((NUMERATE(SORT(DROPMISSING(RESIDUAL)))
      -0.375)/(0.25 + N(RESIDUAL))))) tiplabel=(nq="Quantile" s="Residual")
      tip=(nq s);
endlayout;
```

Again, this code has been manually reformatted. The PROC TEMPLATE SOURCE statement struggles with complicated code like this. Besides having indentation problems, PROC TEMPLATE sometimes breaks lines in the middle of names. These problems must be fixed manually before the generated code can be compiled again by PROC TEMPLATE.

This template differs from others shown previously due to the heavy reliance on expression evaluation. The GTL provides a series of functions that can be used to make plots. The LINEPARM statement produces a diagonal reference line whose slope is the standard deviation of the residuals. A line is determined, given a slope and a point. The Y= option provides the Y coordinate of a point, which is the mean of the residuals. The X= option provides the X coordinate of that same point, which is 0. When X=0, then Y= provides the intercept. The scatter plot consists of the sorted residuals (ignoring missing values) on the Y axis and normal quantiles on the X axis. These quantities are also provided as tooltips. Functions and expressions must always be wrapped in the EVAL function.

Figure 4.19   First BY Group

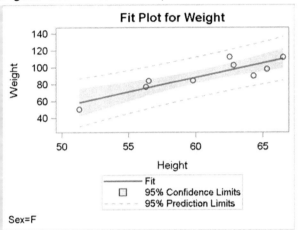

Figure 4.19   First BY Group

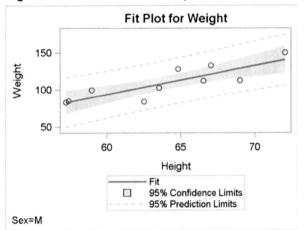

Figure 4.20   Second BY Group

4.9   BY Groups

When a SAS procedure is run with one or more BY variables, SAS provides BY group information at the top of each page of output (for example, "Sex=F" and "Sex=M"). This information does not appear in each table and each graph; it just appears on each page. If you need BY-group information on each graph, you will need to modify the template. For example, you can add the following statements:

```
mvar byline;
entryfootnote halign=left textattrs=graphvaluetext byline;
```

When you add these statements to a template, the value of the macro variable ByLine is displayed as a left aligned footnote. The size, font, and appearance of the footnote comes from the **GraphValueText** style element, which is used for tick and legend labels. Since a macro variable is used, you can modify the template once and reuse it for each BY group. To use the template, you must identify each BY group, extract the BY variable names and values, store them in the macro variable, and run the procedure with just the relevant BY group. You could automate this with a macro and a macro DO loop. The **ModTmplt** macro does all of this for you.

The following steps run PROC GLM with a BY statement and produce separate fit plots for each group:

```
proc sort data=sashelp.class out=class;
   by sex;
run;

proc glm data=class;
   model weight = height;
   by sex;
   ods select fitplot;
run;
```

The graphical results of this step are not shown, but the trace output for the first BY group is as follows:

```
Output Added:
-------------
Name:       FitPlot
Label:      Fit Plot
Template:   Stat.GLM.Graphics.Fit
Path:       GLM.ByGroup1.ANOVA.Weight.FitPlot
-------------
```

The following step displays the template for the fit plot:

```
proc template;
    source Stat.GLM.Graphics.Fit;
run;
```

The results are as follows:

```
link Stat.GLM.Graphics.Fit to Common.Zreg.Graphics.Fit;
```

Some templates are simply links to other templates. The GLM template is a link to a common template that is shared by other procedures. You have two choices. You can copy the common template, change the name to match the GLM template, and then modify it. You can instead modify the common template. This example uses the former approach. The following step displays the common template:

```
proc template;
    source Common.Zreg.Graphics.Fit;
run;
```

A portion of the results are as follows:

```
define statgraph Common.Zreg.Graphics.Fit;
    dynamic _PREDLABEL _CONFLABEL _DEPLABEL _DEPNAME _INDLABEL
        _SHORTINDLABEL _OBSNUM _Y _XVAR _UCL _LCL _UCLM _LCLM _TITLE
        _Y_OBS _XVAR_OBS _PREDICTED _UCL_OBS _LCL_OBS _UCLM_OBS
        _LCLM_OBS _FREQ _WEIGHT _ID1 _ID2 _ID3 _ID4 _ID5;
    BeginGraph;
        entrytitle _TITLE " for " _DEPNAME;
        . . .
    EndGraph;
end;
```

You can create the GLM template as follows:

```
proc template;
   define statgraph Stat.GLM.Graphics.Fit;
       dynamic _PREDLABEL _CONFLABEL _DEPLABEL _DEPNAME _INDLABEL
           _SHORTINDLABEL _OBSNUM _Y _XVAR _UCL _LCL _UCLM _LCLM _TITLE
           _Y_OBS _XVAR_OBS _PREDICTED _UCL_OBS _LCL_OBS _UCLM_OBS
           _LCLM_OBS _FREQ _WEIGHT _ID1 _ID2 _ID3 _ID4 _ID5;
       BeginGraph;
           entrytitle _TITLE " for " _DEPNAME;
           . . .
       EndGraph;
   end;
run;
```

The following steps add an MVAR and ENTRYFOOTNOTE statement to the template and run PROC GLM twice, once for each BY group:

```
%macro mygraph;
   proc glm data=__bydata;
       model weight = height;
       ods select fitplot;
%mend;

%modtmplt(by=sex, data=class, template=Stat.GLM.Graphics.Fit)
```

The results are displayed in Figure 4.19 and Figure 4.20. The ModTmplt macro requires you to provide a SAS macro called MyGraph that contains the SAS procedure that needs to be run. Notice that the BY and RUN statements are *not* specified in the MyGraph macro. The ModTmplt macro constructs and provides them. Also notice that you must use the DATA=__BYDATA option (the data set name begins with two underscores) with the procedure call in the MyGraph macro and specify the real input data set in the DATA= option of the ModTmplt macro. The ModTmplt macro constructs the __BYDATA data set with information from one BY group at a time. If you have more than one BY variable, specify the additional BY variables in the BY= option. You do not need to change anything if you have more BY groups.

You could display the BY group information in a title instead of a footnote as follows:

```
%modtmplt(by=sex, data=class, statement=entrytitle,
          template=Stat.GLM.Graphics.Fit)
```

The results of this step are not displayed. This step adds a second title line with the BY information. Other macro options are available. See Kuhfeld (2009) for more information and options. The macro displays the BY line in a footnote rather than in a title, because most graphs already have titles and few have footnotes. The BY line is less obtrusive in a footnote than it is in a title.

4.10 References

Kuhfeld, W. F. (2009), "Modifying ODS Statistical Graphics Templates in SAS 9.2," http://support.sas.com/rnd/app/papers/modtmplt.pdf.

Appendix A

Introduction to ODS Graphics

Contents

You invoke ODS Graphics by specifying the following statement:

```
ods graphics on;
```

ODS Graphics remains in effect for all procedure steps until you turn it off with the following statement:

```
ods graphics off;
```

Once you have invoked ODS Graphics, creating graphical output with procedures is as simple as creating tabular output. You can control your output as follows:

- ODS destination statements (such as ODS HTML or ODS RTF) specify where you want your graphs displayed.

- ODS SELECT and ODS EXCLUDE statements select and exclude graphs from your output.

- ODS OUTPUT statements create SAS data sets from the data object used to make the plot.

- Procedure options specify which graphs to create. For each procedure, these options are described in the Syntax section of the procedure chapter. Typically, you use the PLOTS= option to control all graphs. The available graphs are listed in the ODS Graphics section, which is found in the Details section of each procedure chapter. Many graphs are produced by default.

- ODS styles control the general appearance and consistency of all graphs and tables.

- ODS templates control the layout and details of each graph.

A.1   ODS Destinations

ODS can send your graphs and tables to a number of different destinations including RTF (rich text format), HTML (hypertext markup language), LISTING (the SAS LISTING destination), DOCUMENT (the ODS document), and PDF (portable document format). You use an ODS statement to open a destination, as in the following examples:

```
ods html body='b.htm';
ods rtf;
ods listing;
ods document name=MyDoc(write);
ods pdf file="contour.pdf";
```

You can close destinations individually or all at once, as in the following examples:

```
ods html close;
ods rtf close;
ods listing close;
ods document close;
ods pdf close;
ods _all_ close;
```

For most ODS destinations (for example, HTML, RTF, and PDF), graphs and tables are integrated in the output, and you view your output with an appropriate viewer, such as a Web browser for HTML. However, the default LISTING destination is different. If you are using the LISTING destination in the SAS windowing environment, you view your graphs individually by clicking the graph icons in the Results window. This action invokes a host-dependent graph viewer (for example, Microsoft Photo Editor on Windows). The graphs produced with ODS Graphics are *not* displayed with traditional graphs in the Graph window.

If you are using the SAS windowing environment and you prefer to view integrated output, you should specify a destination such as HTML or RTF. At the same time, you can prevent the Output window from appearing by closing the LISTING destination, as in the following statements:

```
ods listing close;
ods html;
```

A graph is created for every open destination. When you open a new destination, you should close all destinations that you do not need. Closing destinations makes your jobs run faster and with fewer resources, because fewer graphs are produced.

A.2   Accessing Individual Graphs

If you are writing a paper or creating a presentation, you need to access your graphs individually. There are various ways to do this, depending on the ODS destination. Three particularly useful methods are as follows:

- If you are viewing RTF output, you can simply copy and paste your graphs from the viewer into a Microsoft Word document or a Microsoft PowerPoint slide.

- If you are viewing HTML output, you can copy and paste your graphs from the viewer, or you can right-click the graph and save it to a file. Copying and pasting from RTF is preferable because the default resolution is higher than with HTML.

- You can save your graphs in image files and then include them into a paper or presentation. For example, you can save your graphs as PNG files and include them into a paper that you are writing with LaTeX or into an HTML document.

You can specify the graphics image format and the file name in the ODS GRAPHICS statement. For example, the following statements, when submitted before a procedure step that produces multiple graphs, save the graphs in PostScript files named *myname.ps*, *myname1.ps*, and so on:

```
ods listing close;
ods latex;
ods graphics on / imagefmt=ps imagename='myname';
```

If you are using the LISTING destination and the SAS windowing environment, you can also copy from the default viewer into a Microsoft Word document or a Microsoft PowerPoint slide.

A.3 Specifying the Size and Resolution of Graphs

Two factors to consider when you are creating graphs for a paper or presentation are the size of the graph and its resolution. You can specify the size of a graph in the ODS GRAPHICS statement. The following examples show typical ways to change the size of your graphs:

```
ods graphics on / width=6in;
ods graphics on / height=4in;
ods graphics on / width=4.5in height=3.5in;
```

You can change the resolution with the IMAGE_DPI= option in any ODS destination statement, as in the following example:

```
ods html image_dpi=300;
```

The default resolution of graphs created with the HTML and LISTING destinations is 100 DPI (dots per inch), whereas the default with the RTF destination is 200 DPI. An increase in resolution often improves the quality of the graphs, but it also increases the size of the image file.

A.4 PLOTS= Option

Each statistical procedure that produces ODS Graphics has a PLOTS= option that is used to select graphs and specify some options. The syntax of the PLOTS= option is as follows:

PLOTS < (*global-plot-options*) > < = *plot-request* < (*options*) > >
PLOTS < (*global-plot-options*) > < = (*plot-request* < (*options*) > < ... *plot-request* < (*options*) > >) >

The PLOTS= option has a common overall syntax for all statistical procedures, but the specific global plot options, plot requests, and plot options vary across procedures. There are only a limited number of things that you can control with the PLOTS= option. Most graphical details are controlled either by graph templates or by styles.

The PLOTS= option usually is specified in the PROC statement. However, for some procedures, certain analyses and hence certain plots can appear only if an additional statement is specified. These procedures often have a PLOTS= option in that other statement.

The simplest PLOTS= option specifications are of the form PLOTS=*plot-request* or PLOTS=(*plot-requests*). When there is more than one plot request, the plot-request list must appear in parentheses. Each plot request either requests a plot (for example, RESIDUALS) or provides you with a place to specify plot-specific options (for example, DIAGNOSTICS(UNPACK)). Some simple and typical plot requests are explained next:

- PLOTS=ALL requests all plots that are relevant to the analysis.

- PLOTS=NONE disables ODS Graphics for just that step.

- PLOTS=RESIDUALS requests a plot of residuals in a modeling procedure such as PROC REG.

- PLOTS=RESIDUALS(SMOOTH) requests the residuals plot along with a smooth fit function.

- PLOTS=(TRACE AUTOCORR) requests trace and autocorrelation plots in procedures with Bayesian analysis options.

Global plot options appear in parentheses after the option name and before the equal sign. These options affect many or all of the plots. The UNPACK option is a commonly used global plot option. It specifies that plots that are normally produced with multiple plots per panel (or "packed") should be unpacked and appear in multiple panels with one plot in each panel. The specification PLOTS(UNPACK)=(*plot-requests*) unpacks all paneled plots. The UNPACK option is also used as an option in a plot request when you only want to unpack certain panels. For example, the option `plots=(diagnostics(unpack) partial predictions)` unpacks just the diagnostics panel. In some cases, unpacked plots contain additional information that is not found in the smaller packed versions. The UNPACK option is not available for all plot requests; it is just available with plots that have multiple panels by default.

Another commonly used global plot option is the ONLY option. Many procedures produce default plots, and additional plots can be requested in the PLOTS= option. Specifying PLOTS=(*plot-requests*) while omitting the default plots does not prevent the default plots from being produced. The ONLY option is used when you only want to see the plots specifically listed in the plot-request list. Procedures that produce no default plots typically do not provide an ONLY option. You can use ODS SELECT and ODS EXCLUDE to select and exclude graphs, but in some situations the ONLY option is more convenient. It is typically more efficient to select plots by using the PLOTS(ONLY)= option, because the procedure does not do extra work to generate a plot that is excluded by the PLOTS(ONLY)= option. In contrast, ODS SELECT and ODS EXCLUDE have their effect after the procedure has done the work to generate the plot.

A.5 Viewing Your Graphs in the SAS Windowing Environment

The mechanism for viewing graphics created with ODS can vary depending on your operating system, which viewers are installed on your computer, and the ODS destination you have selected. If you do not specify an ODS destination, then the LISTING destination is used by default. If you are using the SAS windowing environment, go to the Results window and find the icon for the corresponding graph. You can double-click the graph icon to display the graph in the default viewer that is configured on your computer for the corresponding image file type.

If you are using the SAS windowing environment and you specify an HTML destination, then the results are displayed by default in the SAS Results window as they are being generated. Depending on your configuration, this statement can also apply to the PDF and RTF destinations. If you are using the LATEX or the PS destinations, you must use a PostScript viewer, such as GSview. For information about the windowing environment in a different operating system, see the SAS Companion for that operating system.

If you do not want to view the results as they are being generated, then select **Tools ▶ Options ▶ Preference** from the menu at the top of the main SAS window. Then, in the **Results** tab, you can clear the **View results as they are generated** checkbox.

A.6 Determining Graph Names and Labels

Procedures assign a name to each graph they create with ODS Graphics. This enables you to refer to ODS graphs in the same way that you refer to ODS tables. You can determine the names of graphs in several ways:

- You can look up graph names in the ODS Graphics section of chapters for procedures that use ODS Graphics.

- You can use the Results window to view the names of ODS graphs created in your SAS session.

- You can use the ODS TRACE ON statement to list the names of graphs created by your SAS session. This statement adds identifying information in the SAS log (or optionally in the SAS LISTING) for each graph that is produced.

The graph name is not the same as the name of the image file that contains the graph.

A.7 The Default Template Stores and the ODS PATH

SAS stores certain types of information in data sets called item stores. Templates that have been compiled by PROC TEMPLATE are stored in item stores called template stores. The default table, graph, and style templates that SAS provides are stored in the template store Sashelp.Tmplmst. By default, if you modify a template, it is stored in the template store Sasuser.Templat. By default, ODS searches Sasuser.Templat for templates, and then it searches Sashelp.Tmplmst if it does not find the requested template in Sasuser.Templat. You can see the list of template stores by submitting the following statement:

```
ods path show;
```

Here are the results:

```
Current ODS PATH list is:

1. SASUSER.TEMPLAT(UPDATE)
2. SASHELP.TMPLMST(READ)
```

With this default path, SAS always finds the templates that it provides in Sashelp.Tmplmst unless you stored a modified template in Sasuser.Templat. In that case, it finds and uses your modified template in Sasuser.Templat. To see a list of all of the templates that you have modified, submit the following statements:

```
proc template;
    list / store=sasuser.templat;
run;
```

You can delete any template that you modified (so that ODS finds the default template that SAS supplied) by specifying it on a DELETE statement, as in the following example:

```
proc template;
    delete Stat.REG.Graphics.ResidualPlot;
run;
```

ODS never deletes the template in Sashelp.Tmplmst, so you can safely run the preceding step, even if the template you specify does not exist in Sasuser.Templat. You can run the following step to delete the entire Sasuser.Templat template store of customized templates so that ODS uses only the SAS supplied templates:

```
ods path sashelp.tmplmst(read);
proc datasets library=sasuser;
    delete templat(memtype=itemstor);
run;
ods path reset;
```

You can restore the default template path in either of the following equivalent ways:

```
ods path sasuser.templat(update) sashelp.tmplmst(read);
ods path reset;
```

It is good practice to delete templates that you have customized when you are done with them, so that they are not unexpectedly used later.

You can modify the path and insert a Work item store in front of the default path in either of the following equivalent ways:

```
ods path work.templat(update) sasuser.templat(update) sashelp.tmplmst(read);
ods path (prepend) work.templat(update);
```

You can see the list of template item stores by submitting the following statement:

```
ods path show;
```

The results are as follows:

```
Current ODS PATH list is:

1. WORK.TEMPLAT(UPDATE)
2. SASUSER.TEMPLAT(UPDATE)
3. SASHELP.TMPLMST(READ)
```

With this path, any template that you submit is stored in Work.Templat, which is deleted at the end of your SAS session.

The following statements illustrate how you can specify a permanent item store for your use and for the use of others:

```
libname mytpl 'C:\MyTemplateLibrary';
ods path (prepend) mytpl.templat(update);
```

Now, when you run PROC TEMPLATE, SAS creates an item store in the directory you specified in the LIBNAME statement.

A.8 Modifying Your Graphs

Although ODS Graphics is designed to automate the creation of high-quality statistical graphics, on occasion you might need to modify your graphs. There are two ways you can make modifications, depending on whether the changes you want to make are data-dependent and immediate (for a specific graph you are preparing for a paper or presentation), or whether they are persistent (applied to a graph each time you run the procedure). You can make immediate, ad hoc changes by using the ODS Graphics Editor, which provides a point-and-click interface. You can make persistent changes by modifying the ODS graph template for a particular plot. A graph template is a program, written in the Graph Template Language (GTL), that specifies the layout and details of a graph.

You can use the ODS Graphics Editor to customize titles and labels, annotate data points, add text, and change the properties of graph elements. After you have modified your graph, you can save it as a PNG image file or as an SGE file, which preserves the editing context. You can open SGE files with the ODS Graphics Editor and resume editing.

You can invoke the ODS Graphics Editor in the SAS windowing environment, provided that the LISTING destination is open and that you have enabled ODS Graphics to create editable graphs.

A.9   Data Objects

Graphs are constructed from a data component (or data object), a graph template, and a style template. Procedures supply a table of data values and statistical results to plot. Together, the data object and the templates form an output object that ODS displays in one or more output destinations. You can use the ODS OUTPUT statement to create a SAS data set from the underlying data object. You can use PROC PRINT and PROC CONTENTS to display the data set and its variables to learn about the data object and its rows and columns.

A.10   Recommended Reading

More information about ODS, ODS Graphics, the GTL, and SAS/STAT software is available on the Web at:

```
http://support.sas.com/documentation/
http://support.sas.com/documentation/onlinedoc/base/index.html
http://support.sas.com/documentation/onlinedoc/graph/index.html
http://support.sas.com/documentation/onlinedoc/stat/index.html
```

To learn more about ODS, see "Using the Output Delivery System" (*SAS/STAT User's Guide*).

To learn more about ODS Graphics, see "Statistical Graphics Using ODS" (*SAS/STAT User's Guide*).

For introductory information about ODS Graphics, see the documentation section "A Primer on ODS Statistical Graphics" (*SAS/STAT User's Guide*).

For complete ODS documentation, see the *SAS Output Delivery System: User's Guide*.

For complete GTL documentation, see the *SAS/GRAPH: Graph Template Language User's Guide* and *SAS/GRAPH Template Language Reference*.

For complete documentation about the Graphics Editor, see the *SAS/GRAPH: ODS Graphics Editor User's Guide*.

For information about the statistical graphics procedures, see the *SAS/GRAPH: Statistical Graphics Procedures Guide*.

Appendix B

Some Tips and Techniques for Understanding the GTL

The Graph Template Language provides a powerful and general syntax for creating graphs. However, if you are an experienced SAS programmer, you will find that some of the habits and insights that you have developed and that have served you well with other parts of SAS will not transfer to the GTL as well as you might expect. Understanding this will help you later when you write your own templates and things are not working as you expect. The following examples illustrate.

The next steps use PROC TEMPLATE, the GTL, SG procedures, and PROC KDE. We will begin with the briefest of overviews of the GTL. Most of the GTL and procedure syntax is explained in detail later. For now, do not worry about understanding the full syntax. The critical parts are explained as needed and the rest can wait. The following template, written in the GTL, illustrates the basic components of a fully functional graph template:

```
proc template;
   define statgraph Stat.KDE.Graphics.ScatterPlot;
      BeginGraph;
         layout Overlay;
            ScatterPlot x=X y=Y;
         EndLayout;
      EndGraph;
   end;
run;
```

Graph template definitions in PROC TEMPLATE begin with a DEFINE STATGRAPH statement and end with an END statement. Embedded in every graph template is a BEGINGRAPH/ENDGRAPH block, and embedded in that block are one or more LAYOUT blocks. Each layout contains one or more statements that create a graph. The BEGINGRAPH/ENDGRAPH block can also contain entry titles and entry footnotes.

Now, to better understand the differences between the GTL and other SAS syntax, consider the following step that specifies a nonexistent variable:

```
proc print data=sashelp.class;
   var MyMadeUpVariable;
run;
```

When you run this step, SAS produces the following error message:

```
ERROR: Variable MYMADEUPVARIABLE not found.
```

Also consider the following steps, which also specify a nonexistent variable (or to be precise, a nonexistent column in the underlying table or "data object"):

```
proc template;
   define statgraph Stat.KDE.Graphics.ScatterPlot;
      BeginGraph;
         layout Overlay;
            ScatterPlot x=X y=Y;
            SeriesPlot  x=MyMadeUpVariable y=Y;
         EndLayout;
      EndGraph;
   end;
run;

proc kde data=sashelp.class;
   bivar height weight / plots=scatter;
run;
```

PROC KDE is a SAS/STAT procedure that performs kernel density estimation and also produces scatter plots of two data object columns named X and Y that it creates from the input data. It uses the template `Stat.KDE.Graphics.ScatterPlot`. The SERIESPLOT statement was added to the template that SAS provides. The template definition for the series plot specifies a column named MyMadeUpVariable that does not exist. No error messages are produced. First, no error can be issued when PROC TEMPLATE runs, because PROC TEMPLATE cannot see the data object to know if x, y, and MyMadeUpVariable exist. Second, no error is issued by PROC KDE or ODS Graphics. This is by design. No error is issued, because there is nothing wrong with this template. It is perfectly legitimate for templates to specify a value that is nonexistent (for another example, see page 178). When this happens, that part of the graph simply drops out without a note, error, or warning. The template language is designed for maximum flexibility and versatility. This template can be used to make a scatter plot (when both x and y exist), a series plot (when MyMadeUpVariable and y exist), or both (when x, MyMadeUpVariable and y exist). No conditional logic is needed; it all happens automatically and by design. However, there is a down side. When you misspell a name, you might not get a message that helps you figure out what went wrong, because as far as SAS, ODS, and the GTL are concerned, nothing went wrong.

It is important that the GTL work this way. SAS procedures use graph templates for every set of data, reasonable and unreasonable, that gets input. If a template needed extra statements and logic to handle every unreasonable data set or option combination that got thrown at it, it would be much more complicated. Instead, variables that do not exist, are all missing, or are otherwise invalid drop out, a reasonable graph is produced, and the procedure runs quietly without displaying confusing messages about what happened. The GTL works correctly, and it works reasonably, but it might not work the way you expect it to when you make a mistake.

Now consider the following steps, which again specify a nonexistent data object column:

```
proc template;
   define statgraph ScatterPlot;
      BeginGraph;
         layout Overlay;
            ScatterPlot x=height y=weight;
            SeriesPlot  x=MyMadeUpVariable y=weight;
         EndLayout;
      EndGraph;
   end;
run;


proc sgrender data=sashelp.class template=scatterplot;
run;
```

The PROC SGRENDER step produces the following warning:

```
WARNING: The SeriesPlot statement will not be drawn because
         one or more of the required arguments were not supplied.
```

PROC SGRENDER is saying that it does not have one or more of the required arguments (the X= and Y= options) and so the series plot is not drawn. Both options are specified, but one was specified with a nonexistent column, so the specification drops out as if it had not been specified at all. You are probably wondering why SAS prints nothing for the first case and a warning for this second case. The GTL is the language that SAS procedure writers use to produce graphs. When a template is used by an analytical procedure, ODS Graphics assumes that a nonexistent specification is by design, and so the procedure runs quietly, dropping the nonexistent part. However, when PROC SGRENDER is used, ODS Graphics assumes that a user-written template is being used, and so it prints more detailed error messages. Understanding this distinction can be a big help to you when you write and modify templates. Particularly when you modify a template that SAS provides, you cannot rely solely on warnings and error messages to help you understand what went wrong when a graph does not come out right. Rather, you must rely on a thorough understanding of the GTL and how it works. When you write your own template for use with PROC SGRENDER, you will typically get more helpful messages, but you still can't solely rely on SAS messages.

The GTL differs from other SAS syntax in other ways as well. Most SAS procedures have two types of options. Keyword options have the form *keyword=value*. Examples include DATA=*SAS-data-set*, ORDER=DATA, and PLOTS=ALL. The other type of option consists of an option name but no equal sign or value. These options are Boolean (true, false) in nature. By default, the option is not specified; you can specify the option to negate the default behavior. For example, by default, PROC PRINT does not display variable labels, but you can specify the LABELS option to see labels. By default, PROC MEANS displays its output, but you can specify the NOPRINT option to suppress this display.

In contrast, the GTL uses only keyword options: *keyword=value*. Even Boolean options are keyword options and have the form *keyword*=TRUE|FALSE. For example, the option for specifying whether or not a plot is primary is controlled by specifying PRIMARY=TRUE or PRIMARY=FALSE. Some of these options have a default of true, and some have a default of false.

The GTL also does not accept lists like other SAS syntax, and the syntax in the GTL is sometimes verbose. For example, the legacy SAS/GRAPH statement `axis1 order=(0 to 50 by 10)` specifies tick values of 0, 10, 20, 30, 40, and 50. In the GTL, this option is specified as follows: `xaxisopts=(linearopts=(tickvaluesequence=(start=0 end=50 increment=10)))`. However, while the syntax can be verbose, ODS Graphics and the SG procedures do so much for you that ODS Graphics template and programs are often shorter and easier than legacy SAS/GRAPH programs.

You cannot in the GTL specify a statement with a list like the following: `dynamic x1-x6 y1-y6`. Your options include a manual specification (`dynamic x1 x2 x3 x4 x5 x6 y1 y2 y3 y4 y5 y6`) or using the macro language (`dynamic %macro l; %do i = 1 %to 6; x&i y&i %end; %mend; %l`).

The templates that SAS provides are sometimes written using the SAS macro language. For example, the following is a subset of a template that SAS provides, displayed in the way it was originally written with a SAS macro:

```
define statgraph Stat.Transreg.Graphics.PBSplineCriterion;
   notes "Penalized B-spline (CV,GCV,AIC,AICC,SBC) * Lambda Plots";
   %macro plot;
   dynamic %do i = 1 %to 6; s&i %end; nplots nrows ncols head height;
   begingraph / designheight=height;
      entrytitle head;
      layout lattice / columns=ncols rows=nrows order=packed;
         %do i = 1 %to 6;
            if(&i <= nplots)
               layout overlay;
                  scatterplot y = y&i x = x&i / primary=true tip=(y x)
                        tiplabel=(x="Lambda");
                  entry {Unicode Lambda} " = " s&i /
                        autoalign=(topright topleft bottomright bottomleft
                                   right left top bottom);
               endlayout;
            endif;
         %end;
      endlayout;
   endgraph;
   %mend; %plot;
end;
run;
```

This template creates a lattice of up to six plots, arranged in a 2 by 3 grid. Each plot is defined by a LAYOUT OVERLAY block with a SCATTERPLOT statement to make a plot and an ENTRY statement that displays a smoothing parameter in the plot as inset text. Notice that a macro do loop is used to expand the original variable list. (The unabridged template had more dynamic variables and used the statement: `dynamic %do i = 1 %to 6; s&i l&i t&i %end; ....`) Also, a macro do loop is used to generate six overlay layouts, one for x1, y1, and s1 through one for x6, y6, and s6.

The syntax that SAS shows you when you display this template definition is as follows:

```
define statgraph Stat.Transreg.Graphics.PBSplineCriterion;
    notes "Penalized B-spline (CV,GCV,AIC,AICC,SBC) * Lambda Plots";
    dynamic s1 s2 s3 s4 s5 s6 nplots nrows ncols head height;
    begingraph / designheight=height;
        entrytitle HEAD;
        layout lattice / columns=NCOLS rows=NROWS order=packed;
            if (1 <= NPLOTS)
                layout overlay;
                    scatterplot y=Y1 x=X1 / primary=true tip=(y x) tiplabel=(x="Lambda");
                    entry { Unicode LAMBDA } " = " S1 / autoalign=(topright topleft
                        bottomright bottomleft right left top bottom);
                endlayout;
            endif;
            if (2 <= NPLOTS)
                layout overlay;
                    scatterplot y=Y2 x=X2 / primary=true tip=(y x) tiplabel=(x="Lambda");
                    entry { Unicode LAMBDA } " = " S2 / autoalign=(topright topleft
                        bottomright bottomleft right left top bottom);
                endlayout;
            endif;
            if (3 <= NPLOTS)
                layout overlay;
                    scatterplot y=Y3 x=X3 / primary=true tip=(y x) tiplabel=(x="Lambda");
                    entry { Unicode LAMBDA } " = " S3 / autoalign=(topright topleft
                        bottomright bottomleft right left top bottom);
                endlayout;
            endif;
            if (4 <= NPLOTS)
                layout overlay;
                    scatterplot y=Y4 x=X4 / primary=true tip=(y x) tiplabel=(x="Lambda");
                    entry { Unicode LAMBDA } " = " S4 / autoalign=(topright topleft
                        bottomright bottomleft right left top bottom);
                endlayout;
            endif;
            if (5 <= NPLOTS)
                layout overlay;
                    scatterplot y=Y5 x=X5 / primary=true tip=(y x) tiplabel=(x="Lambda");
                    entry { Unicode LAMBDA } " = " S5 / autoalign=(topright topleft
                        bottomright bottomleft right left top bottom);
                endlayout;
            endif;
            if (6 <= NPLOTS)
                layout overlay;
                    scatterplot y=Y6 x=X6 / primary=true tip=(y x) tiplabel=(x="Lambda");
                    entry { Unicode LAMBDA } " = " S6 / autoalign=(topright topleft
                        bottomright bottomleft right left top bottom);
                endlayout;
            endif;
        endlayout;
    endgraph;
end;
```

See the section "Changing Titles and Axis Labels" on page 159 for an example of displaying template syntax.

Before PROC TEMPLATE compiles the template, the macro language expands the variable lists and produces the six overlay blocks. Then PROC TEMPLATE compiles the template and stores the results. Then you use PROC TEMPLATE to reproduce the template source from the compiled template, not from the original source. This process can change spacing, remove comments, and change capitalization. When you look at templates that SAS provides, it is important to understand that they might have looked very different when they were developed. Some templates that SAS provides are much longer than this one. However, they are usually much less complicated than they might appear at first glance. While this version of the template is certainly longer than the original, macro version, it has the same straightforward organization of a lattice with six layouts that each define a scatter plot.

It is important to understand that the GTL is a language developed for internal use at SAS. It was developed to be a comprehensive language for capturing the definition of a potentially very complex graph. While SAS makes this language available to you, it is not designed to be an obvious step away from a syntax that is familiar to long-time SAS users. It is a new and different language. That said, you will find as you become experienced with the GTL that it is not hard to use. It is just in some ways different than you might expect if you are an experienced SAS user.

Index

LaVergne, TN USA
08 December 2010

207823LV00001B/4/P